舌尖上的
广西“非遗”

The Taste of Guangxi Intangible Cultural Heritage

宫 斐 编著

广西民族出版社
Gvangjsih Minzcuz Cuzbanjse

图书在版编目(CIP)数据

舌尖上的广西“非遗”/宫斐编著.--南宁:广西民族出版社,2024.10.--ISBN 978-7-5363-7791-2

Ⅰ.TS971.202.67

中国国家版本馆CIP数据核字第20242LX329号

SHEJIAN SHANG DE GUANGXI “FEI-YI”

舌尖上的广西“非遗”

出 版 人:石朝雄
策划编辑:王博书 梁秋芬 梁 倩 何世春
责任编辑:梁 倩
美术编辑:何世春
责任校对:庞丽明 黄毅萍 孙 书 白雨思
装帧设计:何世春
责任印制:张东杰
出版发行:广西民族出版社
地址:广西南宁市青秀区桂春路3号 邮编:530028
电话:0771-5523216 传真:0771-5523225
电子邮箱:bws@gxmzbook.com
印 刷:广西壮族自治区地质印刷厂
规 格:240毫米×170毫米 1/16
印 张:16.5
字 数:255千
版 次:2024年10月第1版
印 次:2024年10月第1次印刷
书 号:ISBN 978-7-5363-7791-2
定 价:128.00元

《舌尖上的广西“非遗”》编委会

主　　任：罗生芳

副 主 任：郭　勇　许晓明

委　　员：刘　铨　李　宁　邵　威　宫　斐

英文审订：李　宁　谭艳红

指导单位：

广西壮族自治区文化和旅游厅非物质文化遗产保护处

广西壮族自治区非物质文化遗产保护中心

序 Preface

杨铭铎

非物质文化遗产（以下简称“非遗”）是人类的特殊遗产。中华饮食文化博大精深、历史悠久，拥有很多经典的“非遗”饮食。这些“非遗”饮食要求选择特色食材，使用不同的烹调技法，最终呈现出五味调和、百般品相风味。“非遗”饮食承载着当地的历史记忆，蕴含着厚重的饮食文化。

经历了时间的更替和打磨，进入现代叙事语境，对“非遗”饮食的挖掘、记录及其制作技艺的传承与发展，具有非常重要的价值。

一是文化价值。“非遗”饮食是中华文化的重要组成部分。研究“非遗”饮食，可以深入了解不同地区和民族的文化传统与历史渊源，对于保护和传承中华文化具有重要意义。

二是社会价值。“一方水土养一方人。”“非遗”饮食是当地社会交往的重要媒介，能让人们感受不同地方、不同群体的饮食文化差异，丰富人们的生活体验，促进文化的交流，促进社会融合。

三是经济价值。随着生活水平的提高，人们对于传统美食的需求也在不断增加。通过对“非遗”饮食的挖掘和推广，可以带动相关产业的发展，如餐饮业、旅游业等，从而促进地方经济发展。

四是健康价值。“非遗”饮食蕴含着丰富的养生理念，如“养助益充”、药食同源等理念，对于现代人的健康养生具有重要的指导作用。

广西的饮食丰富多彩，地方特色浓郁，如近年来火到海外的柳州螺蛳粉、飘香海内外解乡愁的桂林米粉、被誉为“中国一绝”的梧州纸包鸡等，都具有独特的地域风味。

感谢作者，让我先睹为快，了解广西的饮食地图和饮食特色。《舌尖上的广西“非遗”》一书对于广西具有代表性的“非遗”饮食的介绍，不仅有历史渊源的追溯，有对传承人的采访，有对制作技艺传承的梳理，还配有精美的图片和视频，并对关键内容进行英文翻译，内容全面、系统，极具创新和传播价值。该书既可以作为“非遗”饮食文化传承和发展的研究资料，又可以作为大众了解和认知“非遗”饮食的通俗读本，做到了雅俗共赏，值得一读。

“民以食为天。”“非遗”饮食既是我们日常生活的一部分，也是人类多彩文化的重要组成部分。在现代叙事语境下，“非遗”饮食文化的保护与传承需要与时俱进，审慎处理好表达通俗性与准确性的关系，大胆创新，倡导表达方式的多样性，提升触达途径的有效性。

该书系广西“非遗”饮食研究成果，是为序。

作者简介：杨铭铎，中国烹饪协会餐饮教育工作委员会主席，中国食文化研究会副会长，中国烹饪协会非物质文化遗产事业推进指导委员会专家。

目录 Contents

食之地图：一方水土　一方饮食

Food Map: the unique features of a local environment always give special characteristics to its foods

食之技艺：技知此味艺业精

Food Craftsmanship: mastery of technique brings out the essence of taste

◆ 菜肴类

Dishes

68 / 全州醋血鸭：醋血为魂　家乡至味

Shelduck in Vinegar Blood, Quanzhou Style: vinegar-blood as the soul, the most authentic local flavor

◆ 米粉类

Rice Noodles

78 / 南宁老友粉：热辣酸爽　有情有义

Nanning Laoyou Rice Noodles: spicy and sour, full of heart and soul

90 / 柳州螺蛳粉：汤鲜笋臭　酸辣爽口

Liuzhou Luosifen: fresh soup and fermented bamboo shoots, spicy and sour, refreshing to the palate

106 / 桂林米粉：卤水鲜香　米粉滑韧

Guilin Rice Noodles: rich and fragrant sauce, smooth and chewy noodles

121 / 南宁生榨米粉：软糯香滑　开胃健脾

Nanning Squeezed Rice Noodles: soft, glutinous, fragrant, and smooth, appetizing and good for the spleen

◆ 茶酒类
Tea and Liquor

◆ 小吃类
Snacks

◆ 其他
Others

食之文化：“那”边风味独好
Food Culture: "Na" has a unique flavor

舌尖上的广西『非遗』

食之地图：一方水土 一方饮食

Food Map: the unique features of a local environment always give special characteristics to its foods

饮食是民族文化的载体，也是地域特色的延续。不同于川菜的麻辣、鲁菜的咸鲜、粤菜的鲜香，广西饮食似乎很难用一个词概括。纪录片《舌尖上的中国》的导演陈晓卿就曾评价“广西是中国饮食最复杂的地方”。

一方水土，一方饮食。广西地处中国南疆，东连广东，南临北部湾并与海南隔海相望，西与云南毗邻，东北接湖南，西北靠贵州，西南与越南接壤，居住着壮、汉、瑶、苗、侗、仫佬、毛南、回、京、彝、水、仡佬等 12 个世居民族。“八山一水一分田”的山地丘陵性盆地地貌、亚热带季风气候及热带季风气候并存的气候特征，多民族交融聚居的人文环境，形成了广西丰富多彩的饮食文化和风俗。

俯瞰八桂大地，从合浦县山口镇的英罗港到防城

The unique features of a local environment always give special characteristics to its foods. Guangxi is located in the southern border of China, west of Guangdong, north of Beibu Gulf and Hainan across the sea, east of Yunnan, southwest of Hunan, southeast of Guizhou, northeast of Vietnam. It has been inhabited by 12 ethnic groups, including Zhuang, Han, Yao, Miao, Dong, Mulao, Maonan, Hui, Jing, Yi, Shui, and Gelao. The unique geographical, climate and cultural environment have formed a rich and colorful food culture and customs in the region.

港市东兴市的北仑河口，从华南第一高峰猫儿山到中国四大渔场之一的北部湾渔场，广袤的山、河、湖、海、田，孕育了丰富多样、绿色天然的物产资源，禽畜种类繁多，蔬果四时不断。优质的食材，追求传统与天然的饮食理念，加上各具特色的制作方法，形成了广西“东甜西酸、南鲜北辣”的地域饮食格局。2021 年 2 月 1 日施行的《桂菜标准体系》将之进一步丰富，概括为桂北辣咸、桂西酸辣、桂东南甜酸、滨海鲜咸的风味特征。其中，桂北风味主要指桂林、柳州、来宾、贺州一带的饮食风味，桂西风味主要指百色、河池和崇左一带的饮食风味，桂东南风味主要指南宁、梧州、玉林、贵港一带的饮食风味，滨海风味主要指钦州、北海、防城港一带的饮食风味。

With the use of high-quality ingredients, the pursuit of traditional and natural food concepts, as well as the adoption of unique and ingenious cooking techniques, Guangxi has formed a regional diet pattern characterized by "sweet in the east, sour in the west, fresh in the south, and spicy in the north". The *Guangxi Cuisine Standard System* implemented on February 1, 2021 further enriched it, which is summarized as spicy and salty in northern Guangxi, sour and spicy in western Guangxi, sweet and sour in southeastern Guangxi, and fresh and salty in the coastal area.

桂北风味：融合湘楚特色　自成一方美食

Northern Guangxi Flavor: a unique cuisine with Hunan and Hubei characteristics

鸭肉为形、醋血为魂的全州醋血鸭是桂北名菜，看似『黑暗料理』，食则酸香开胃，让人欲罢不能。

秦代灵渠的修建，连通了湘江和漓江，打通了中原与岭南的交通要道，使得中原文化、湘楚文化较早传入桂北地区，并与本土文化碰撞交流，形成“湘桂走廊文化带”。明朝至民国时期，大量湖广籍移民涌入桂北一带，使得以桂林为中心的各个城镇基本流行西南官话，饮食也深受湘楚风味影响，口味偏辣、咸，非常注重调料，喜欢干香入味的菜肴，尤其是桂柳地区以辣著称，民间曾流传有“兴全灌，没有辣椒不送饭”之说。主要的名菜、名小吃有桂林米粉、柳州螺蛳粉、恭城油茶、全州醋血鸭、荔浦芋扣肉和阳朔啤酒鱼等。

The diet of the northern region of Guangxi is deeply influenced by the Hunan and Hubei flavors, with a salty, fragrant, and spicy taste. It emphasizes on seasoning, especially in Guilin and Liuzhou regions. The dishes are known for their spiciness. There has been a saying that "People can't eat without chili peppers in Xingan, Quanzhou and Guanyang (3 places in northern Guangxi)". The main famous dishes and snacks are Guilin Rice Noodles, Liuzhou Luosifen, Gongcheng Oil Tea, Shelduck in Vinegar Blood, Quanzhou Style, Steamed Sliced Pork with Lipu Taro, and Yangshuo Beer Fish.

桂北地区饮食口味醇厚，色泽浓重，无辣不欢；很多经典的传统菜肴制作讲究，工艺繁复，独具匠心。桂北地区有名的全州醋血鸭正是其中之一，其以盛夏上市的仔鸭为主料，杀鸭留血，在鸭血中注入酸醋，以嫩姜或苦瓜等为配料，在出锅前倒入醋血而成，做法独树一帜，口味酸辣鲜香，让人百吃不厌。桂北名菜阳朔啤酒鱼亦是一绝，以啤酒入菜，很好地除去鱼的土腥味，使鱼肉更加细腻、鲜辣可口，让人食之难忘。

善炖扣肉亦是桂北地区的一大饮食特色。桂北地区有着“无扣肉不成席”的习俗。荔浦芋扣肉是荔浦乃至桂北地区的传统名菜，其将芋头与五花肉完美结合，吃起来醇香酥烂、肥而不腻。桂北名菜——松皮扣肉工序复杂、烹制过程考究，先将肥瘦相间的肉煮透，用牙签在猪皮上扎些细密的小孔，再炸至猪皮起泡，过凉水后用料腌制，最后蒸煮而成。其肉皮弹韧，肉质鲜美，被陈晓卿导演赞誉为“要闭着眼睛咽下去”的一道美味。

桂北地区春夏多雨潮湿，秋冬阴寒酷冷，早上起来喝一碗油茶是当地人的习惯。尤其在恭城，人们一年四

In the northern Guangxi region, it is rainy and humid in spring and summer, while gloomy and cold in autumn and winter. Drinking a bowl of Oil Tea in the morning is a habit of local people. Especially in Gongcheng, Oil Tea can be consumed throughout the year for three meals a day. Oil Tea is meticulously crafted by slowly pounding tea leaves and ginger together, which as the main ingredients. Long-term consumption of Oil Tea is believed to have various health benefits, such as dispelling cold and dampness, improving digestion and promoting stomach health.

季、一日三餐都离不开油茶。油茶是以茶叶、生姜为主料慢慢捶打而成的，长期饮用油茶，有驱寒祛湿、健胃消食等功效，近年来被越来越多的人所喜欢。此外，灌阳、融水、兴安、三江等地也流传着打油茶的饮食风俗，制法上大同小异，配料上各有不同，如灌阳油茶除了有茶叶和生姜之外，还有绿豆、香菇，甚至有排骨、猪血、瘦肉和青菜等，作料里除了有炒米、葱花、花生之外，还有酸辣椒，这是灌阳油茶的灵魂所在。

在广西，生津开胃的酸是必不可少的一道口味。桂北地区的酸味有其独特之处。例如，近年来柳州螺蛳粉凭借酸、辣、鲜、爽、烫的独特风味频频“出圈”。酸汤猪脚也是桂北地区家喻户晓的一道名菜。将猪脚先炖后炸，随后放入酸笋、酸辣椒、酸姜一起炖，吃起来鲜香浓郁、酸辣爽口。

桂北地区是最早连通中原与岭南的区域，其饮食风格独特，具有广阔的发展前景，柳州螺蛳粉的持续火爆就是桂北饮食走向全国的成功案例。在“舌尖文化”的推动下，桂北饮食将受到更多关注。

The northern Guangxi region is the first area to be connected to the Central Plains and Lingnan, boasting a unique dietary style that has vast development prospects. With the promotion of food culture, the northern Guangxi cuisine will receive more attention.

桂西风味：天然食材丰富　长寿文化加持

Western Guangxi Flavor: rich in natural ingredients with longevity culture blessing

巴马烤香猪选用巴马当地的小香猪，配上各种香料炭烤，香气四溢，可谓『七里之遥能闻其味』。

桂西地区位于广西西部，山环水绕，资源丰富，居住着壮、瑶、苗、回、彝、仡佬等多个少数民族，其中壮族约占总人口的 80%。

桂西地区的饮食具有浓厚的民族风味，擅长众菜合调，粗菜细做。各少数民族利用当地土特产制作出本民族喜欢的风味菜，尤其对江河中的野生鱼种，高山蔬菜，山间野菌，田埂野菜，乡村土鸡、土鸭情有独钟，又特别强调食材季节性、成熟期，突出菜肴味鲜、味纯的特点，口味上偏向酸辣。主要名菜、名小吃有瑶乡竹筒饭、百色脆皮狗、巴马烤香猪、环江香牛扣及西林羊瘪汤等。

The cuisine of the western Guangxi region has a strong ethnic flavor, with a good blend of various dishes and fine cooking of coarse dishes. Ethnic minorities cook their favorite specialty dishes with local specialties, especially wild fishes in the rivers, wild vegetables from fields, and local chickens and ducks. These dishes are fresh and authentic, with a preference for salty, sour, and spicy flavors.

桂西地区山清水秀，孕育了丰富的食材，使得当地饮食呈现出就地取材，集美味、营养、保健于一体的特色。以“世界长寿之乡”巴马瑶族自治县为例，其特殊的饮食结构被认为是影响巴马人寿命最直接、最明显的因素。巴马人常年以玉米、稻米等为主食，常吃豆类食物，大量食用蔬菜，常吃火麻仁……巴马人由此形成了“五低两高”(即低热量、低脂肪、低动物蛋白、低盐、低糖，高维生素、高纤维素）的健康饮食特色。

喜食蔬菜，尤其是野生蔬菜，是桂西地区的饮食特色。相关研究显示，桂西地区壮族饮食文化中常用的野生食用植物有 46 科 102 种，其中菊科种类最多（13 种)，其次为苋科、伞形科、蔷薇科和豆科（各 6 种）等。野生食用植物的采摘时间主要为春季和夏季，部分种类可全年采食，食用方法有炒、煮、凉拌等。

在肉类饮食方面，桂西地区极具民族风味。壮族人擅长用动物的血制作美食，如用猪血做成的血肠就是当地的一种特色美食。杀猪时，用盆装盐水接血，不停搅拌使新鲜的猪血不凝固，拌上碎肉和优质糯米，佐以葱蒜等，拌匀后将其灌入洗净的猪肠中，用细线扎成一节节，入锅煮熟即可。此外，侗族的竹笋肉，苗族的竹板鱼、羊瘪汤等都是当地的特色美食。

The western region of Guangxi has picturesque mountains and rivers, beautiful environment and abundant ingredients. The local diet presents a unique feature of locally sourced ingredients, integrating delicacy, nutrition, and health care. People like eating vegetables, especially wild vegetables, which is a dietary characteristic in the western Guangxi region. In terms of meat diet, people in the western Guangxi region also adapt to local conditions, adopting multiple varieties and refined techniques, extremely ethnic flavor.

桂西地区大多数民族都有食酸的饮食风俗。苗族的酸食以红白酸最为经典：白酸汤用淘米水或者米汤酿造，糯米水更佳；红酸汤是将西红柿放入坛子中并加入盐、酒、姜等发酵而成，用其制成的酸汤鱼尤其开胃。扶绥壮族酸粥是将煮好的新鲜白米饭舀出放凉后，倒入干净的坛子中，并加入少许酸粥益生菌种，在正常温度内经过一周左右的时间发酵腌制而成。在崇左，酸粥不仅是主食，还是像盐一样不可或缺的作料，当地人用它和不同的食材进行搭配，形成了独具特色、营养丰富的酸粥宴。

丰富多彩的民族文化在相互交融中形成了桂西地区独具特色、自成体系的饮食习俗，制法独特、讲究实惠、富有山野乡土气息是桂西地区的饮食特色。尤其在天然食材及长寿文化的加持下，桂西地区的饮食文化在新时期将得到更广泛的传播。

Most ethnic groups in western Guangxi have a habit of eating with a slightly sour and salty taste. The Miao ethnic group's sour diet is characterized by red or white sour soup being the most classic. In Chongzuo, sour congee is not only a staple food, but also an indispensable seasoning like salt.

桂东南风味：清淡鲜嫩口感　食疗养生内核

Southeast Guangxi Flavor: light, fresh and tender taste with the core of dietary therapy and health care

被誉为『中国一绝』的梧州纸包鸡，观之色泽金黄，入口鲜嫩甘滑，食后齿颊留香。

自古以来，桂东南地区就是人文荟萃之地。秦汉以来，因从军、流徙、行商、避乱等，中原移民不断迁入广西。桂东南地区自然资源丰富、土地肥沃、气候适宜，尤其是地理位置优越、水路纵横：西江水系沟通了广西大小河流，总汇于梧州；自梧州上溯桂、浔、郁、黔、柳、贺诸江，可贯穿广西各处而达滇、黔、湘三省及越南；东下西江，可通粤港各埠。因此，桂东南地区一直是中原移民迁居广西的主要区域之一，本土文化、中原文化及广府文化在此碰撞、交融，形成独具特色的桂东南文化，属于岭南文化的延伸和发展。

桂东南风味菜用料比较广泛，良禽蔬果皆可入菜，口

Since ancient times, southeastern Guangxi has been called the place of "gathering of talents". Its natural resources are rich, the land is fertile, the climate is suitable, especially the geographical position is superior and the waterways are extensive. The Xijiang River system connects the rivers of Guangxi and converges at Wuzhou. It has always been one of the main migration areas of Han people entering Guangxi. The local culture, Central Plains Culture and Cantonese Culture collide and blend here, forming the unique southeast Guangxi culture.

味以清淡为主，讲究鲜嫩滑爽。当地人喜欢食用生鲜、脆嫩、偏甜的菜肴品种。桂东南地区以粉食为代表的风味小吃更是样多味美。主要名菜、名小吃有梧州纸包鸡、武鸣柠檬鸭、南宁老友粉、玉林牛肉巴、博白白切菜等。

从饮食风格来看，桂东南地区做菜讲究鲜、活、嫩。以肉类菜品为例，梧州纸包鸡被定位为“传统节庆的第一道菜”，充分体现了梧州菜肴选料严格、制作精细、刀工讲究、口味丰富的特点。其采用独特的隔纸浸炸制作技艺，使滚油不入内，味汁不外泄，这种似炸而非炸的制作技艺保留了鸡肉的鲜嫩口感。玉林的白切菜颇具名气，从白切鸡到白切鸭，再到白切猪脚等，讲究原汁原味，味道多清淡鲜美，酱料香醇，深受当地人喜爱。而横县鱼生（注：2010 年横县鱼生制作技艺入选广西第三批自治区级“非遗”代表性项目名录。经国务院批准，2021 年 2 月 3 日，撤销横县，设立县级横州市。然而，人们还是习惯称“横县鱼生”，本书沿用此称）更是以鲜、活、嫩为最大特色，以选材新鲜、刀工精细、摆盘精美、口感丰富闻名于世，完美诠释了“食不厌

The dishes of southeastern Guangxi have a wide range of ingredients, including fine poultry, vegetables, and fruits. They are mainly light in taste, emphasizing freshness, tenderness, and smoothness. People like to eat raw, crisp, tender and sweet dishes. Flavor snacks, represented by rice noodles, are very diverse and delicious. The main famous dishes and snacks include Wuzhou Paper-wrapped Chicken, Wuming Lemon Duck, Nanning Laoyou Rice Noodles, Yulin Dried Beef, Bobai White Cut Series and so on.

精、脍不厌细”的饮食遗风。

另外，桂东南地区气候炎热多雨，湿气较重，所以在饮食上注重清热降火，喜欢以药入膳。在桂东南地区，随处可见的凉茶铺是其城市的特色景观之一。凉茶其实就是用中草药熬成的汤汁，种类繁多，有雷公根水、菊花茶、二十四味茶等，有清凉解毒、败火利喉的功效。而历史悠久、声名在外的传统药膳龟苓膏是以鹰嘴龟和土茯苓为主料，由数十种纯中药材熬炼而成，味道微苦而甘甜，具有清热祛湿、滋阴补肾、养颜提神等功效，充分显示了桂东南地区药食同源的饮食特征。

桂东南地区口味清淡，这里的酸偏向甜酸，比如南宁人离不开的“酸嘢”（方言，意为用果蔬腌制而成的酸品小吃）、宾阳酸粉，玉林人爱吃的酸甜扣肉，梧州人钟爱的牛肠酸，等等。这些酸食酸甜可口，美味回甘，独具地方饮食特色。

总的来说，桂东南地区的饮食受到了百越族的饮食习惯、中原饮食传统及粤菜影响，多种饮食风俗相辅相成，形成独具特色的饮食文化。

The climate in southeast Guangxi is hot and rainy with heavy humidity, so people pay attention to clearing heat and reducing fire in diet and like delicacy with medicinal effect. The taste in southeast Guangxi is light, and the acid here tends to be sweet and sour. In general, the diet of southeast Guangxi includes the eating habits of Baiyue peoples, and inherits the traditional diet of Central Plains and Cantonese area.

滨海风味：味从舌尖过　大海心中留

Coastal Flavor: taste the heart of the sea from the tip of the tongue

沙蟹汁浓缩了大海最原始的滋味，是广西北海的特色作料，最常用于焖豆角。

广西滨海地区是一片物产富饶的区域：钦州是“中国大蚝之乡”；防城港是“中国八角之乡”“中国肉桂之乡”；北海是著名的港口城市，海味干货闻名全国，盛产优质海盐。

由于特殊的历史渊源，广西滨海地区与近代广东同根同源，是桂粤两地文化的交融地，形成了兼具滨海风味及广式风味的饮食文化。其表现为用料广、选料严，擅长以海鲜为主料烹制菜肴，菜品讲究调味，注重配色，口味上偏向鲜咸，追求鲜嫩、爽滑的口感，配菜、粥品及点心特别丰富。主要的名菜、名小吃有沙蟹汁炒豆角、钦州烤大蚝、北海海鲜粉、蒜蓉银丝扇贝、马鲛

Due to its unique historical origins, the coastal areas of southern Guangxi share the same roots and origins with modern Guangdong, forming a culinary culture that combines coastal and Cantonese flavors. It is characterized by a wide range of materials and strict selection. Being good at taking seafood as the main raw material, people pay attention to seasoning and color matching when cooking. The taste highlights fresh. The dishes are light, fresh and smooth, and the side dishes, congee and dim sum are particularly rich.

鱼丸、盐花煎鱼等。

俗话说“靠海吃海”，广西滨海地区食材以海鲜最为突出。钦州出产的鱼、虾、蟹等海产有500多个品种，四大名海产——大蚝、青蟹、石斑鱼、对虾，肉质鲜美、营养丰富。北海的海鲜丰富，鱼、虾、蟹、蛍、螺、蛤等常年不缺。尤其是鱼，种类繁多，是北海人爱吃的主菜，因而有“北海人猫相”之说。防城港亦是海产丰富，最抢手的海货是石斑鱼、鲈鱼、鳗鱼、黄鲷鱼、比目鱼、鲳鱼等，还有品类众多的海虾、蟹、螺类等，深受当地人喜爱。海鲜的烹饪方法有清蒸、白灼、炭烤、焖炒、盐焗、煎炸、煲煮等，不一而足，但以清蒸、白灼最多，讲究“一沸而起、甘鲜脆美”，力求保持海鲜最鲜美的味道。

除此之外，独特的海鲜作料也是该区域的一大饮食特色，传承了原汁原味的古风。鱼露是防城港“京族三岛”的特产，其以各种小杂鱼为原料，经过腌渍、发酵、过滤、晒炼等多道工序制作而成，鲜咸味美，是当地人吃海鲜、荤肉，乃至煮汤、炒菜必不可少的作料。北海沙蟹汁则是广西北海独有的，以自然生长在沿海滩涂上的沙蟹为原料，通过自然发酵的方法制成的一种纯天然特色作料。在当地，无论是蒸、煮、煲、炒，人们都喜欢放入沙蟹汁，这样即便

Seafood is the most prominent food in the coastal areas of southern Guangxi. In terms of cooking methods of seafood, there are various options, including steaming, blanching, charcoal grilling, braising, salt roasting, frying, and pot boiling. However, steaming and blanching are the most common, striving to maintain the flavor of seafood. In addition, the unique seafood dipping ingredients are also a major culinary feature of the region, inheriting the authentic ancient customs of ancestors.

不放一点盐，也能为食物带来鲜咸、清甜的口感。当地的一道特色美食——沙蟹汁焖豆角，就用了沙蟹汁进行调味提鲜，这道菜被评为广西十大经典名菜之一。

由于广西滨海地区常年气候炎热，同时受广州饮食习俗的影响，各类好喝又方便的汤水、粥品是当地人的“心头好”。据说，北海人在外就餐时坐下来说的第一句话常是：“老板有什么汤？”北海人尤其爱喝鱼汤，如石斑鱼汤、杂鱼汤等。同时，粥的种类也很多，如花生粥、清简粥及各种海味粥等，做法随人们的喜好而定。被称为“天下第一粥”的东兴三鲜粥的原料其实不止 3 种海鲜，而是囊括了蛤、沙虫、海鱼、鱿鱼、蟹等各类食材，这样一锅海鲜盛宴，其美味和营养可想而知。

酸是广西各地离不开的饮食口味，由于人们平时食用海鲜比较多，广西滨海地区的酸味也讲究一个“酸鲜”。不少北海人在制作海鲜作料时会加上几滴醋调味。此外，钦州当地的酸制品也很多，较有名气的酸制品有酸黄瓜皮、酸芋蒙、酸菜、酸豆角、酸葡萄和酸萝卜等。其中，钦州黄瓜皮具有粒粒鲜嫩、颗颗香脆的特色，曾与荔枝、龙眼一起被誉为“岭南三宝”，素有“宁舍鱼翅燕窝，不舍钦州瓜皮”之说，如今已远销海内外。

Due to the fact that it is hot for a long time all the year round in the coastal areas of southern Guangxi, and influenced by the dietary customs of Guangzhou, local people are fond of all kinds of tasty and convenient soup and congee. Acid is an indispensable food taste for Guangxi people. Due to the frequent consumption of seafood in daily life, the sour taste in the coastal areas of southern Guangxi also emphasizes a "sour freshness". Many Beihai people add a few drops of vinegar to flavor their seafood juice. In addition, there are also many acidic products in Qinzhou.

味从舌尖过，大海心中留。广西滨海地区以其独具特色的饮食风格及文化，丰富了广西餐饮版图。在北部湾开发、大湾区建设及“一带一路”建设深入推进的背景下，滨海风味美食作为地方饮食名片，展示着当地独特的文化风俗，吸引着游客前来品味打卡，同时在开放发展中博采众长、海纳百川，融入更多来自全国各地和东南亚的味道。

美味的海鲜 ◆广西滨海地区最显眼的美食名片

Delicious seafood ◆The most prominent culinary delight of Guangxi's coastal regions

从武鸣柠檬鸭、梧州纸包鸡到巴马香猪、壮族血肠，从东兴三鲜粥、钦州黄瓜皮到恭城油茶、柳州螺蛳粉……在广西，不同民族、不同地域有着不同口味的特色美食，看似彼此独立，实则相互影响，大道融合。诚如《人间风味》里所说的："美食，总在迥然不同的民族和文化间扮演着亲善大使。静默的岁月中，彼此渗透，新的风味就这样奇妙诞生。"

俗话说："五味调和百味香。"广西桂菜发展促进会会长王卫东认为："桂菜最大的特点是口味丰富，嗜咸、嗜甜、嗜辣、嗜酸的人都能找到可口的美味。"这也正是天南海北的人来到广西都能很快适应并彻底融入的原因所在。

深入了解桂菜，人们更能感受到其包容并蓄又特色鲜明的内核，而"非遗"饮食作为其中极具历史文化及技艺传承价值的代表，将成为人们由表及里去触碰八桂土地广博饮食文化的媒介，溯历史，传技艺，看传承，观发展。

In Guangxi, different ethnic groups in different regions have unique cuisine with different flavors, which are independent but mutually influential. By gaining a deeper understanding, one can better feel the inclusive and unique core of Guangxi cuisine. As the representative boasting extremely high historical, cultural, and artistic inheritance value, intangible cultural heritage cuisine will become a medium for people to touch the extensive culinary culture of the land of Guangxi from the outside to the inside.

舌尖上的

广西『非遗』

食之技艺：技知此味艺业精

Food Craftsmanship: mastery of technique brings out the essence of taste

武鸣柠檬鸭：酸辣香浓　生津开胃

Wuming Lemon Duck: the soul collision of lemon, duck and sauce

"非遗"饮食名片

◆ 武鸣柠檬鸭制作技艺

类别：传统技艺

级别：自治区级"非遗"代表性项目

批次：第七批

地域：广西南宁

酸椒柠檬开胃汁，武鸣柠檬鸭好吃。酸酸辣辣的武鸣柠檬鸭，味浓而冲，鸭肉不软不腻不臊，带有特别的柠檬香气，这样的开胃美食，吃上一次便难以忘怀。

作为南宁本土原汁原味的"草根美食"，武鸣柠檬

With pickled sour peppers, lemons and sauce, Wuming Lemon Duck is delicious. The sour and spicy Lemon Duck has a strong taste. The duck meat is crispy and meaty with a special lemon aroma, and it is not too soft or greasy. Such an appetizer is unforgettable after eating it once. In 2018, Wuming Lemon Duck Making Technique was selected into the seventh list of representative projects of Intangible Cultural Heritage at the autonomous region level in Guangxi.

扫码看视频

生津开胃的柠檬鸭 ◆当鸭子碰上柠檬酸料，便成就了这道极具特色的地方菜肴
Lemon Duck, refreshing and appetizing ◆When duck meets lemony sauce, it creates this distinctive local cuisine

鸭被评为广西十大经典名菜之一。2018 年，武鸣柠檬鸭制作技艺入选广西第七批自治区级“非遗”代表性项目名录。

溯源：因路而起，因路而兴

柠檬鸭发源地是南宁市武鸣区，当地壮族人口约占 86%，有“中国壮乡”之誉。古时，武缘（武鸣旧称）去往邕州（南宁旧称）的必经之路，被高峰挡住。民国三年（1914 年），武鸣籍人士陆荣廷任两广都督。他主政时，劈山开路，修建广西第一条公路——邕武路。

有一天，陆荣廷检查修路情况，村民们宰杀土鸭，用自家腌制的柠檬、酸辣椒、蒜头和生姜等作料烹饪菜肴招待他。陆荣廷品尝后大加赞赏。“陆荣廷界牌吃鸭”的故事不胫而走。

由于当时修建公路，大批民工聚居工地，村民便在路边支起小店。随着道路开通，通往云贵两省的运输车辆和人流不断增多，柠檬鸭的名气渐渐传开，成了南宁的一道特色美食。

武鸣柠檬鸭制作技艺自治区级代表性传承人、甘家界柠檬鸭第三代传人甘保辉回忆，柠檬鸭是他们那边逢年过节、接待客人的一道家常菜。

The birthplace of Lemon Duck is Wuming District, Nanning City. In 1914, Lu Rongting served as the governor of Guangdong and Guangxi. When he was in reign, he hosted to build Yongwu Road, the first highway in Guangxi. The villagers once cooked local ducks with their homemade pickled lemons, sour peppers, garlic, ginger, and other spices to treat him. Lu Rongting tasted and praised it highly. The story of "Lu Rongting Enjoying Duck at the landmark" soon spread far and wide. After that, the villagers set up small shops by the roadside and catered for a large number of migrant road workers. With the opening of the road, the dish of Lemon Duck gradually became a delicacy in Nanning.

20 世纪 70 年代末，甘保辉的父亲甘喜开在高峰林场界碑分场门口的邕武路旁开了第一家甘家界柠檬鸭店（店名是甘家界牌柠檬鸭总店）。那时的邕武路是广西"第一路"，往来车辆很多，甘家界柠檬鸭店主要是供过往司机吃饭和休息的。俗话说："一方水土，一方饮食。"南宁气候炎热，人们在饮食上偏清淡，喜好开胃的酸类食品，柠檬鸭最初是偏酸的。当时过路的多是云南、贵州的司机，他们喜食辣，就建议加些辣椒。甘喜开凭着自己的悟性，在祖辈技艺的基础上加以改进，武鸣柠檬鸭的口味就慢慢地演变为今天的酸辣口味。凭着独特的口味，柠檬鸭声名日增。

随着生意越来越兴隆，来到邕武路吃柠檬鸭的老顾客越来越多。机缘巧合下，柠檬鸭迎来了第一次发展机遇。20 世纪 80 年代末期，时任南宁市市长的甘祥梦先生回武鸣老家时经常去邕武路吃柠檬鸭。出于对家乡美食的喜爱，甘市长建议将甘家界柠檬鸭推广出去，打造成广西名牌菜。

在甘市长的推介及媒体的宣传报道下，甘家界柠檬鸭有了更大的知名度，并靠着口碑传播，被越来越多的南宁市民所喜爱。

20 世纪 90 年代，甘家界柠檬鸭已盛名在外。南宁至武鸣的二级公路开通后，邕武路的车流量不如从前了，但从南宁市区专程

At the end of the 1970s, Mr. Gan Xikai opened his first shop near Yongwu Road mainly for passing drivers to have a meal and rest. On the basis of cooking technique of his ancestors, Gan Xikai improved the Wuming Lemon Duck with his own understanding, and the dish slowly evolved into today's sour and spicy type with the aroma of lemon. With its unique taste, Wuming Lemon Duck began to be recognized again and its reputation grew day by day.

过来的食客却越来越多。那时私家车还比较少，很多人都是坐公交车或者骑摩托车赶过来，就为了吃一顿柠檬鸭。“想吃鸭子到郊外”成了当时的风潮。

由于到邕武路吃柠檬鸭的人太多，店里坐不下，人们就把桌

甘家界牌柠檬鸭总店　◆20 世纪 80 年代深受南宁食客喜爱的“美食打卡点”
Ganjiajie Lemon Duck Main Branch at Gaofeng　◆A popular spot for foodies in Nanning, since the 1980s

椅摆到外面的大树下、停车场里去吃，那场面看上去就像村里的流水席一样热闹。当时的餐饮环境没有现在这么精致，却如邻里聚餐一般有着浓郁的烟火气和人情味。店里的人忙不过来，顾客就帮忙砍鸭子、洗青菜，有些顾客一路相伴走过来，已和甘家人成了老朋友。

在市场需求的激励下，1998 年甘喜开在武鸣开了自家品牌的第二家直营店；2004 年在南宁开了第三家直营店……经过多年发展，“甘家界柠檬鸭”已发展成为连锁品牌。柠檬鸭也伴随着一个年代的记忆，成为南宁人家喻户晓的一道菜肴。

技艺：柠檬、鸭子与酱料的灵魂碰撞

“喝鸭汤，吃鸭肉，一年四季不咳嗽。”鸭子是广西的一种主要食材，尤其是在农历七月十四这天，广西人家家户户都要吃鸭子，由此广西人过的中元节也被戏称为“鸭子节”。同时，广西人嗜酸，可谓“万物皆可酸”。当鸭子碰到柠檬酸料，便成就了柠檬鸭这道极具特色的地方菜肴。

在甘保辉看来，柠檬鸭这道菜的灵魂在于“生态散养的本地鸭 + 传统腌制老柠檬 + 家传秘制酱料”。

Duck is one of the staple diet in Guangxi. On July 14th of the lunar calendar, every household in Guangxi eats duck. Meanwhile, Guangxi people have a strong preference for sour taste, to the extent that "everything can be made with vinegar". When duck meets lemon sauce, it creates the highly distinctive local dish of Lemon Duck. Gan Baohui believes that the soul of the Lemon Duck lies in: ecological free-range local ducks plus traditional pickled lemons plus Gan's secret sauce.

新鲜柠檬 ◆个小皮薄，肉质鲜嫩

Fresh lemon ◆Small size, thin skin, and tender flesh

腌柠檬 ◆腌制 3 年以上才能沉淀出独特的口味

Pickled lemon ◆A three-year fermentation process to attain its unique taste

（1）选材

要做好柠檬鸭这道菜，选材是第一步。制作柠檬鸭，一定要选当地散养的体表光滑、身材匀称的土鸭作为主料。其肉厚肥美，爆炒时很能出油提香，出锅后口感不柴，并有嚼劲。

制作柠檬鸭时，柠檬选用的是南宁当地的土柠檬，皮薄，果肉饱满多汁，腌制 3—7 年的最合适，这种柠檬剁碎了有糖浆流出，香味浓郁；辣椒要选用南宁当地腌制的酸辣椒，肉厚，没有太多水分；姜则要选大的肉姜。

The duck meat of Lemon Duck must be selected from local free-range ducks. The lemons used must be the small lemons native to Nanning with thin skin, full and juicy flesh; those pickled for three to seven years are the most suitable. The duck is slaughtered and cut into chunks to absorb the sauce better. Pit and cut the aged pickled lemons into strips. Cut the sour peppers into pieces, shred the fresh gingers, and prepare the minced garlic. Ginger, garlic, sour peppers and a variety of spices are cooked in an authentic way to make a secret sauce, which is the main determinant of whether Lemon Duck is delicious.

（2）准备

将鸭子宰杀好并切成块状，以利于更好地吸收酱料。将腌制好的柠檬剁碎，酸辣椒切段，生姜切丝，备好蒜头。

将生姜、蒜头、辣椒及多种调料以地道手法烹制而成的秘制酱料，是柠檬鸭好吃的关键。

秘制酱料 ◆柠檬鸭好吃的关键

Secret sauce ◆The key to delicious Lemon Duck

（3）爆炒

在实际烹饪中，让柠檬酸料与鸭肉完美融合在一起，离不开“炒”的成全。将新鲜鸭肉与生姜、蒜头、酸辣椒煸炒后，加适量水焖煮，再加入秘制酱料一起爆炒。待鸭肉半熟，再加入切好的柠檬酸料一起炒，让鸭肉与柠檬酸料充分结合。

在这个过程中，掌握好火候是关键。在猛火爆炒下，生出让人食欲大增的由六分酸爽、三分咸香、一分微辣组成的乡野美味，再用文火焖到鸭肉熟透，接着改用大火翻炒至汁水成糊状。

（4）出锅

鸭的油脂与酱汁要收干得刚好，香而不焦，既入味又不干涩。将制作好的柠檬鸭盛入特制的铁锅或砂煲中，端上桌，在其底部置蜡烛继续加热，油亮金黄的色泽和四溢的柠檬香气令人垂涎欲滴。即使是在盛夏，光闻味道都会让人食欲大增。吃一口，香、酸、辣等滋味层层递进，让人欲罢不能。

传承：地方饮食文化的延续与突围

如今，柠檬鸭在不断传承和优化中，早已成为南宁人心心念念的一道家乡菜，也成为地方饮食文化的典型代表。甘保辉认为，柠

Stir-fry fresh duck meat with ginger, minced garlic, and sour peppers, then stir-fry them by adding the secret sauce. When the duck is half cooked, add the sour lemons and stir-fry to allow the duck meat to fully combine with the lemons. In this process, mastering the heat is the key. First stir-fry over a high heat, then simmer until the duck is cooked through, then switch to a high heat and stir-fry until the juice is into a paste. The duck fat and sauce should be dried up just right to make the meat aromatic but not burnt or not too dry in taste.

色香味俱佳的柠檬鸭 ◆吃上一次，便让人难以忘怀

Delicious and fragrant Lemon Duck ◆Once tasted, it's unforgettable

檬鸭能从路边小店的招牌菜发展成为今天深受大众喜爱的“非遗”饮食，最大的因素是“有味道、有温度、有传承、有融合”。甘家界柠檬鸭也通过餐饮连锁化发展，将这一地方饮食推广到更多地方。

2020 年疫情突至，许多餐饮企业或倒闭，或苦苦挣扎，或另寻出路。一些老牌的“非遗”饮食餐饮企业在突围之路上苦苦求索。

餐饮食品化，成为甘家界的发展选择。甘保辉介绍，早在 2014 年“甘家界”就开启了袋装柠檬鸭的业务：“最开始是因为有些熟识的顾客不在南宁，又想念柠檬鸭的味道，就委托我们将真空包装后的柠檬鸭邮递过去，于是慢慢就开启了真空袋装柠檬鸭的业务。后来这种需求越来越多，再加上袋装螺蛳粉发展火热，我们就开始考虑进行袋装食品的市场布局。”

2017 年，广西甘家食品有限公司建立了能够独立生产、包装、售卖袋装柠檬鸭的完整产业链条；2022 年，取得了食品生产许可证（SC）。但相比餐饮堂食，袋装柠檬鸭产品一直处于不温不火的状态，属于补充性业务。疫情期间，传统线下门店受到重大冲击，预包装产品业务快速增长，成为疫情下企业经营的一道亮光。不论何时何地，只要想吃柠檬鸭，加热即食的袋装柠檬鸭就能为每一位热爱柠

In Gan Baohui's view, the crucial factor in the development of Lemon Duck from a signature dish in a roadside shop to today's popular diet of intangible cultural heritage is "taste, warmth, promotion and integration". Ganjiajie Lemon Duck Restaurant has also promoted this local food to more places through the development of restaurant chains. The sudden arrival of the epidemic has greatly affected traditional offline stores. Under the new retail model, the growth of the bagged Lemon Duck product business has become a bright light for the operation of enterprises under the epidemic.

檬鸭的人还原出地道的味道。

把柠檬鸭这道地方特色美食推广到全国乃至全世界，是甘保辉为之奋斗的最大愿景。目前，甘家界柠檬鸭袋装产品已经在电商平台和新媒体平台等新零售渠道铺开，线上业务近几年也持续增长，其中 60% 的销量集中在两广地区，40% 的销量在国内一线城市。“餐饮实体店 + 新零售”模式，逐步构建了“商品宣传 + 体验 + 购买”的线下及线上复合型渠道，创新了“非遗”饮食餐饮企业的品牌发展路径。

除市场化传承之外，甘保辉近年来还通过开办企业培训班、参与“非遗”宣传交流活动等，传播武鸣柠檬鸭制作技艺。父亲曾对他说：“一辈子做好一件事就好了。”而武鸣柠檬鸭制作技艺的推广与传承，正是他身体力行的毕生事业。

未来，甘保辉将依托自主创建的柠檬鸭制作技艺传承基地的平台优势，增强“非遗”文化宣传的互动性、体验性，让更多人有机会参与到武鸣柠檬鸭制作技艺的保护与传承当中来。

In addition to marketization, Gan Baohui has in recent years disseminated the Wuming Lemon Duck Making Technique by setting up corporate training courses and participating in intangible cultural heritage publicity and exchange activities. In the future, Ganjiajie Lemon Duck will rely on the platform advantages of the self-established Lemon Duck Making Technique Heritage Base to enhance the interaction and experience of intangible cultural heritage publicity, so that more people will have the opportunity to preserve and sustain the Lemon Duck Making Technique.

梧州纸包鸡：鲜嫩甘滑　齿颊留香

Wuzhou Paper-wrapped Chicken: delicately tender, smooth and sweet, leaving an aftertaste in your mouth

"非遗"饮食名片

◆ **梧州纸包鸡制作技艺**

类别：传统技艺

级别：自治区级"非遗"代表性项目

批次：第六批

地域：广西梧州

"食在岭南，不能不尝纸包鸡"，说的就是梧州纸包鸡。其以独特的玉扣纸浸炸烹饪法，锁住鸡肉以及调料原有的味道，做出来的鸡肉鲜嫩甘滑、色泽金黄、醇厚不腻、齿颊留香，被誉为"中国一绝"。2016年，梧州纸包鸡制作技艺入选广西第六批自治区级"非遗"代表性项目名录。

"Eating in Lingnan, you can't help but taste Paper-wrapped Chicken", which refers to Wuzhou Paper-wrapped Chicken. In 2016, Wuzhou Paper-wrapped Chicken Making Technique was selected into the sixth list of representative projects of Intangible Cultural Heritage at the autonomous region level in Guangxi.

扫码看视频

溯源：千年传承的“岭南名片”

“无鸡不成宴”，尤以广西为甚，纸包鸡便是其中一种极具特色的吃法。中国当代著名国学家、文化学者肖健撰文介绍，纸包鸡作为宫廷贡品和日常主菜的历史至少已有 2200 年，其始创于岭南地区。据考证，人们在见证了岭南地区文明史的西汉南越王墓中，就发掘出许多类似于纸包鸡的鸡骨头和残纸。相传，赵佗建立南越国，定都番禺（今广州），当时的南越王宴就是以纸包鸡作为主菜的。公元前 202 年，刘邦称帝，史称“汉高祖”，建立西汉王朝，定都长安（今西安）。赵佗以纸包鸡作为贡品献给刘邦，上表臣属西汉。刘邦品尝了纸包鸡之后龙颜大悦，诏封赵佗为南越王，至此天下归于大一统。民间流传着这样的赋作：“南粤称臣兮，归一统；纸包金鸡兮，定乾坤。”纸包鸡遂成为南越国向西汉王朝年年进贡、岁岁来朝的贡品。

北宋大文豪苏轼曾这样赞叹纸包鸡：“食遍南粤菜，最忆纸包鸡。”清代国学大师康有为称赞纸包鸡：“纸包鸡者，粤菜之祖宗也，天下奇珍，莫能与之匹敌。”

"No chicken, no banquet", especially in Guangxi, Wuzhou Paper-wrapped Chicken is one of the most distinctive cooking methods. Xiao Jian, a famous contemporary ancient Chinese civilization and national culture scholar, wrote an article introducing that paper-wrapped chicken has been used as a court tribute and daily main dish for at least 2,200 years, and was founded in the Lingnan region.

纸包鸡 ◆纸包不住火，但包得住鸡的鲜香

Paper-wrapped Chicken ◆Paper cannot wrap fire, but it can enclose the fresh and savory aroma of chicken

近代伟大的民主革命先行者孙中山先生称赞纸包鸡："此鸡其声名已响彻粤桂之地也。"

据肖健考证，西汉初期以来，纸包鸡在制作技艺上，一直都是使用酱油与五香粉对鸡肉进行调味腌制的，用于包裹鸡肉的纸张通常是普通草纸；到了民国初期，在岭南美食文化的另一个重要发祥地——广西梧州，纸包鸡的制作技艺获得了重大突破。

据史料记载，从清代咸丰年间开始，纸包鸡就被列为梧州府府台宴请宾客的主菜。民国初年，梧州北山脚下同园"环翠楼"的掌厨师傅在选材、配料、刀工、调味与火候等方面，对纸包鸡的制作工艺做了全面提升，使之色、香、味更佳，酒楼声名鹊起。

民国五年（1916 年），粤西楼在梧州创立，聘请粤菜名厨官良坐镇。官良经反复实践，独创出用白酒、八角、茴香与草果等 20 多种香料制成的纸包鸡秘制卤汁，同时将包裹纸包鸡鸡肉的纸张由原来的普通草纸改为以清香的嫩竹和清澈的山泉水为原料所制作的玉扣纸，这种纸包鸡制作技艺一直沿用至今。粤西楼的纸包鸡作为其招牌菜，名扬岭南。

据说，在 20 世纪 30 年代的某天，广东军政要员陈济棠在广州设宴款待宾客，特地派手下驾驶飞机到梧州购买粤西楼的纸包鸡，

According to historical records, since the Xianfeng period of the Qing Dynasty, Paper-wrapped Chicken has been listed as the main dish of official banquets in Wuzhou Mansion. In the fifth year of the Republic of China (1916 AD), Yue Xilou was founded in Wuzhou, employing the famous Cantonese chef Guan Liang to manage. Guan Liang created a secret marinade made of more than 20 kinds of spices such as white liquor, star anise, fennel and grass fruits. At the same time, the paper used to wrap chicken was changed from ordinary grass paper to Yukou paper has been used so far.

以及特制酥皮狗和原汁三蛇酒。隔天“陈济棠派飞机到梧州粤西楼买纸包鸡”成为热点新闻，粤西楼的纸包鸡更是名声大振，由此民间也有了“饮食在梧州，鸡飞狗酒属一流”的笑谑之谈。

在时代变迁下，纸包鸡几度沉浮。1983 年，在北京人民大会堂举办的全国烹饪美食大赛上，梧州名厨陈北水先生制作的鲜美嫩滑的纸包鸡引得人人争相品尝，名扬京华，赢得了“中国一绝”的美誉。1985 年，纸包鸡还被选拍入《中国一绝》风情电视纪录片，其以独特的制作技艺和品质口感，再次惊艳四方。

如今，随着“舌尖经济”的崛起，纸包鸡作为梧州的招牌美食，在历史传承与时代创新的融合下，进入了产业化发展的新征程。

技艺：隔纸浸炸的独特制法

制作纸包鸡，一般以当地农家散养的三黄鸡作为原料，切件后，用秘制卤汁腌制，以玉扣纸逐件包裹，再放入锅中用花生油浸炸而成。作为一道具有 2000 多年历史底蕴的美味佳肴，纸包鸡充分体现了梧州菜肴选料严格、制作精细、刀工讲究、口味丰富的特点。

梧州纸包鸡制作技艺自治区级代表性传承人、粤菜名厨黄永林一直致力于纸包鸡制作技艺的传承和弘扬。在他看来，纸包鸡与其

The traditional Paper-wrapped Chicken is generally made by selecting local free-range Sanhuang chickens as raw materials, cutting the chicken into pieces, marinating them with secret marinade, wrapping them one by one with Yukou paper, and then deep frying them in a wok with peanut oil. As a delicacy with a history of more than 2,000 years, Paper-wrapped Chicken fully reflects the characteristics of Wuzhou cuisine with strict selection of ingredients, fine production, exquisite knife work and rich taste.

他菜肴相比，最大的特色在于汁多味浓、色黄肉香。

（1）选材

首先，要优选正宗的梧州当地农家散养的三黄鸡（要选未下过蛋的母鸡，当地称其为“项鸡”）。所谓“三黄鸡”，是因羽黄、喙黄、爪黄而被朱元璋钦赐美名的土鸡品种，其肉质细嫩，味道鲜美，营养丰富，在国内外享有较高的声誉。

其次，要选用玉扣纸。这是一种传统手工造纸，曾是皇室贡品。玉扣纸以清香的嫩竹和清澈的山泉水为原料制作而成，不仅纸质柔韧、耐油浸、炸时不易脆烂，还带有淡淡的天然植物香味，是制作纸包鸡的好材料。

最后，秘制卤汁是纸包鸡独特味道的主要来源。传统纸包鸡的卤汁主要由生抽、草菇老抽、白糖、味精、白酒、五香粉（用八角、花椒、桂皮、丁香、小茴香磨成的粉）、胡椒粉、麻油、花生油及姜汁等制成。纸包鸡的制法基本大同小异，只不过卤汁有少许出入，有加甘草粉的，有加少许蜜糖或蚝油的，因而形成了各家的不同特色。

（2）炸纸

取上好的玉扣纸若干张，裁成25厘米左右见方大小，再放入

In terms of material selection, firstly, the authentic farmhouse Sanhuang chicken free-ranged outdoors is preferred. Its meat is tender, delicious and nutritious. Secondly, select the Yukou paper which is a kind of traditional handmade paper. It was once a royal tribute. Made of fragrant tender bamboo and clear mountain spring water, it is flexible, resistant to oil immersion, and not easy to break when fried with a faint natural plant aroma. Finally, the sauce is the main source of the unique flavor of Paper-wrapped Chicken.

70℃的生油锅内炸 2 分钟左右，要注意炸透而不能炸糊，捞起后用手撕不易碎。

（3）包鸡

一只 1.5 千克左右的三黄鸡，只选取肉质最好的 6 个部位（两个翅膀、两个鸡腿和两个背部），将其切成一条连刀不断的梳子形，分成大小均等的若干件（每件重约 75 克），放入提前调制好的秘制卤汁中，腌半小时左右，使之入味。再将鸡肉置于玉扣纸之上，撒些葱粒，包成“日”字形（注意封口要紧密）。

（4）隔纸浸炸

独特的隔纸浸炸烹饪法，是制作纸包鸡的关键所在。其最大的特点在于工艺上是浸炸，但鸡肉又是非炸的，主要是通过油温去浸熟。把包裹好的纸包鸡放入 180—200℃的油锅中，然后将油锅拿出来，离灶炸 6—7 分钟，等油温凉到 100℃左右，再将油锅入灶，猛火炸 1—2 分钟，待鸡肉夹杂着秘制卤汁的香味出来后，即可出锅。

在浸炸的过程中，调控温度是关键。纸包鸡刚入锅高温油炸后要及时离灶，否则纸容易炸黑，同时要注意翻边，让肉浸炸均匀。目前有些店铺采用恒温控制的油炸方式制作纸包鸡，将其放

Take a few pieces of Yukou paper, cut them into 25 cm squares, and then put them into a 70℃ raw oil wok and fry them for about 2 minutes. Take a Sanhuang chicken of about 1. 5 kilograms, and only choose the six parts with the best meat quality (two wings, two legs, two backs), put them into the pre-prepared sauce, and marinate. Then place the chicken on the Yukou paper, and wrap them. Put the wrapped chicken into the oil wok, Soak and fry the meat wrapped by paper. Take the meat out of the wok as soon as the aroma of the chicken comes out.

纸包鸡 ◆玉纸轻挑满堂香

Paper-wrapped Chicken ◆Lightly unwrapped, its aroma fills the room

在 170℃的油锅中浸炸 9 分钟左右即可，但炸出来的味道不如用传统的调温方法制作出来的。

纸包不住火，但包得住鸡的鲜香。在玉扣纸的保护下，滚油不入内，味汁不外泄，独特的烹饪技艺使鸡肉的汁水被玉扣纸牢牢锁住，让鸡肉保留鲜香而不油腻。

“玉纸轻挑满堂香”，熟透的纸包鸡香气四溢，吃时不用特地把纸解开，用筷子轻轻划开一个“十”字，金黄诱人的鸡肉便显露出来，原汁原味的鸡汤清晰可见，趁热品尝，满口留香。难怪朱元璋会这样赞叹：“此鸡只应天上有，人间能得几回尝。”

纸包鸡不仅口感好，营养价值也十分高。三黄鸡散养时间在 150 天以上，肉质达到了最佳状态，细嫩鲜美、蛋白质含量高，易消化、易吸收，并且用秘制卤汁腌制，药食同源，有强身健体之功效。

传承：“活态流变”的发展创新

经过代代师传，如今纸包鸡已成为梧州的传统名菜和美食名片。凡来梧州的中外游客和商人，无不以能品尝到纸包鸡为一大快事。

With the Yukou paper, boiling oil does not seep into and the flavor juice does not leak out. By the unique cooking technique, the juice of the chicken is firmly locked by the Yukou paper, so that the chicken retains its fresh flavor and abundant nutrition without being greasy. The free-range breeding period of Sanhuang chicken is more than 150 days, so its meat quality has reached the best. The meat is tender and delicious, containing a high level of protein easy to digest and absorb. Because it has marinated with Chinese herbal ingredients, it has the effect of strengthening the body.

近年来，梧州市政府有关部门与企业共建“梧州纸包鸡制作技艺传习所”，实施代表性传承人带徒授艺制度，组织传承“非遗”文化的大型活动，收集、管理、存档与纸包鸡制作技艺相关的历史资料和实物等。此外，还坚持组织纸包鸡进校园、进社区等活动，让纸包鸡制作技艺和文化深入大众。

在传承古老技艺的同时，纸包鸡从业者也在谋新谋变，以更好地适应时代需求。黄永林经过长时间的研究摸索，将材料的用量统一标准化后，利用现代生产设备制作纸包鸡，并倡导研发系列产品，如黑椒纸包鸡、香辣纸包鸡、咖喱纸包鸡等。他认为，只有通过多样化口味的研发，才能更适应现代消费者个性化的喜好需求，有利

纸包鸡 ◆独特的隔纸浸炸技艺，让鸡肉保留鲜香而不油腻

Paper-wrapped Chicken ◆Unique frying technique with paper wrapping preserves the freshness and aroma of the chicken while preventing greasiness

于纸包鸡这道名菜走出梧州，被更多人喜欢。

为保障纸包鸡出品的标准性、稳定性，梧州纸包鸡制作技艺自治区级代表性传承人林艳光与梧州饮食业商会共同制定了纸包鸡产品的企业标准。在此基础上，林艳光通过工艺创新，将纸包鸡的油摄入量减少 60%，以契合现代人“吃出健康”的理念。

同时，林艳光首创保鲜便携装纸包鸡，创新性地解决了纸包鸡不能保鲜、不便携带的难题，使得纸包鸡这道传统美食突破地域限制，成为“广西手信”，并通过线上及线下渠道，走向更广阔的市场，推动纸包鸡向产业化、品牌化、全国化发展。

林艳光还对纸包鸡制作技艺进行创新，使纸包鸡进入“酱烤时代”。不同于传统的隔纸浸炸烹饪法，酱烤烹饪法是将切好的鸡肉用秘制酱料腌制后，放进烤炉里进行高温酱烤，待酱烤完毕，再用炸好的玉扣纸包裹上，然后进行真空保鲜包装和高温灭菌。在高温灭菌的过程中，玉扣纸的香味完全渗透到鸡肉中。目前，这两种制作技艺均在梧州并存。

欲成其事，先扬其名。在纸包鸡创变革新的过程中，建设纸包鸡“文化高地”使得纸包鸡这道传统美食重获高人气、高热度的做法，亦值得关注和学习。2012—2015 年，林艳光与中国当代著名国

After a long period of research and exploration, Huang Yonglin standardized the amount of materials, then he advocated the research and development of a series of products, such as Black Pepper Paper-wrapped Chicken. Lin Yanguang reduced the intake of oil of Wuzhou Paper-wrapped Chicken by 60% through process innovation. He also pioneered the portable fresh-keeping Paper-wrapped Chicken, and further innovated the production technology, so that Wuzhou Paper-wrapped Chicken entered the "sauce-roasted era".

学家、文化学者肖健强强联手，积极倡导弘扬纸包鸡文化，通过新华网、中国网等中央重点新闻网站发表多篇关于纸包鸡文化定位的学术论文，溯源考证纸包鸡历史、严谨挖掘纸包鸡文化，将其定位为“传统节庆的第一道菜”，大力推动纸包鸡文化的传播。如今，纸包鸡的文化定位已在海内外形成广泛共识，深入人心。肖健对于纸包鸡“活态流变的特性”的阐释也广受认可，重塑了纸包鸡文化的当代价值。

传承老味道，变革新工艺；溯源旧历史，发掘新内核。纸包鸡在传承发展中，由一道餐桌美食，变为地方“手信”，如今又成为梧州美食文化产业发展的引领者，在“活态流变”中，继续焕发着新的活力。

With the process of innovation and reformation of Wuzhou Paper-wrapped Chicken, the construction of a "cultural high ground" which has made the traditional delicacy regain its high popularity is also noteworthy, reshaping the contemporary value of Paper-wrapped Chicken culture. The evolution and development of Wuzhou Paper-wrapped Chicken has transformed a table delicacy into a Wuzhou souvenir, and now Paper-wrapped Chicken has become the leader in developing Wuzhou food culture industry, continuing to release new vitality in the "dynamic evolution".

横县鱼生：脍不厌细　圣人遗风

Hengxian Sliced Raw Fish: saint legacy of "Sliced Raw Fish can't be more thin"

"非遗" 饮食名片

◆ **横县鱼生制作技艺**

类别：传统技艺

级别：自治区级 "非遗" 代表性项目

批次：第三批

地域：广西南宁

"透如碧玉薄如翼，鱼脍爽口好滋味。"横县鱼生是著名的地方传统佳肴，其以选材新鲜、刀工精细、摆盘精美、口感丰富闻名于世。2010 年，横县鱼生制作技艺入选广西第三批自治区级 "非遗" 代表性项目名录。

Hengxian Sliced Raw Fish is a renowned traditional local delicacy that is famous for its fresh ingredients, meticulous slicing skill, elegant plating, and rich flavor. In 2010, Hengxian Sliced Raw Fish Making Technique was selected into the third list of representative projects of Intangible Cultural Heritage at the autonomous region level in Guangxi.

扫码看视频

横县鱼生　◆透如碧玉薄如翼

Hengxian Sliced Raw Fish　◆As transparent as the jasper, as thin as the feather

溯源：古老淡水鱼脍的“活化石”

提起鱼生（也称“生鱼片”，俗称“两片”），人们大多以为其是来自日本的舶来品，殊不知鱼生的吃法起源于中国，有着悠久的历史。

早在原始居民定居于河边和海边后，人类便有了食鱼历史。鱼生在我国古代又称“鱼脍”“脍”，因为做脍的原料以鱼为多，所以“脍”又写作“鲙”。《礼记》有“肉腥细者为脍”的记载，《汉书》中明确指出“生肉为脍”。

我国最早的关于吃鱼生的文字记录，可追溯到先秦时期，如《诗经》所记载的“饮御诸友，炰鳖脍鲤”中的“脍鲤”即生鲤鱼片；春秋时期，《论语》中有孔圣人主张“食不厌精，脍不厌细”的记载；战国时期，《孟子》中有关于“脍炙人口”的记载；到了唐代，食脍盛行，不少诗词中都有它的留影，如李白的《拟古十二首》中的“吹箫舞彩凤，酌醴鲙神鱼”，杜甫的《夜二首》中的“暂忆江东鲙，兼怀雪下船”，白居易的《轻肥》中则提及“脍切天池鳞”，等等；鱼生在宋元时期的热度依然不减；然而在明清时期，这道菜肴逐渐淡出人们的视野；直至今日，古老的生鱼脍习俗仅在两广等少数区域顽强地延续了下来，堪称古老淡水鱼脍的“活化石”。

The original inhabitants settled in rivers and seashores, and humans have a long history of eating fish. The earliest written record of eating sliced raw fish in China can be traced back to the pre-Qin period. In the Tang Dynasty, eating sliced raw fish was prevalent, and it appeared in many poems. The popularity of sliced raw fish in the Song and Yuan Dynasties remain unabated. However, In the Ming and Qing Dynasties, this dish began to gradually fade out of people's vision. Until today, the ancient custom of eating sliced raw fish has only tenaciously passed down in a few areas such as Guangdong and Guangxi.

广西横州吃鱼生的习俗古已有之。据清代《横州志》记载，最先吃鱼生的是几千年前生活在郁江两岸的横州先民——疍人。《横州志》记载了一个民间传说，相传晋元帝时，逸士董京避世于横州。一日，董京携疍人兄妹泛舟赏月，忽见仙人乘槎而来。董京请客，仙人称不饮熟食，不辞而别。疍人兄妹进山寻究，在山洞偶遇仙人，仙人以玉液琼浆和鱼生款待。后来，兄妹俩按仙人指点制作鱼生。从此，横县鱼生名声大振，流传至今。

横州吃鱼生的习俗，有文字记载的可追溯至明朝。明朝的徐霞客在游历广西地区时，就发现当地“取巨鱼细切为脍，置大碗中，以葱及姜丝与盐、醋拌而食之，以为至味”。明代进士周孟中在《登春野亭》一诗中以“匙翻玉粒长腰米，脍切银丝缩项鳊”赞叹横县鱼生形色之精妙。清代《广西通志》也有记载：“剖活鱼细切，备辛香、蔬、醋诸物下箸拌食，曰‘鱼生’，胜于烹者。”

横县鱼生鲜美，得益于一条穿境而过的大河——郁江，其水流清澈，水域开阔，鱼苗麇集，所产特盛。秦观在《宁浦书事》中曾赞叹“鱼稻有如淮右，溪山宛类江南”，古来就聚集着舟楫为家、捕鱼为业的水上人家。古时渔民在郁江上捕鱼为生，由于没有调味品，也不能生火，捕到的鱼只能生吃，后来几经演变和改良，鱼生成了

Hengzhou in Guangxi has been eating raw fish from ancient times. The written record of eating raw fish in Hengzhou can be traced back to the Ming Dynasty. When Xu Xiake of the Ming Dynasty traveled to Guangxi, he found that the local people "took giant fish finely cut into sliced raw fish, mixing it with green onions, sliced ginger, salt and vinegar in a large bowl, and tasted it". In the Qing Dynasty's *Guangxi Tongzhi*, there is also recordation: "Cut live fish finely, prepare spices, vegetables and vinegar and mix them. The dish is known as 'sliced raw fish' which is better than cooked fish. "

横州的第一美食。

技艺：飞刀生花 “鱼”味无穷

横县鱼生之所以闻名遐迩，与其精挑细选的食材和独特的制法息息相关。横县鱼生制作技艺有着严格的要求。在横州，只有将每片鱼片控制在厚度 3 毫米、重量 8—10 克，配料切得细如发丝且大小均匀的厨师，才能称为真正的鱼生师傅。

横县鱼生制作技艺自治区级代表性传承人余富曾在广东做厨师，出于对家乡菜品的热爱以及传承的使命感，他放弃了广东的高薪，回到家乡学习制作鱼生。如今，余富投身鱼生餐饮行业已有 20 多年，有着“横县鱼生第一刀”的名号。据他介绍，制作鱼生需要经过以下 5 个步骤：养鱼、放血、取肉去皮、切片和装盘。

（1）养鱼

制作鱼生最好选用郁江鲜美的河鱼，如野生鲩鱼、花鱼、鲮鱼和青竹鱼等，同时必须保证鱼的生猛、新鲜及洁净。余富介绍，通常野生鱼的体形稍小，如果鱼肚比鱼头还要大的就不是野生鱼。其鱼鳞呈青黄色，鱼尾和鱼背的肉质较为结实。

捕捞上来的鱼不可立即用于制作鱼生，须要净养数日，余富将

The reason why Hengxian Sliced Raw Fish is famous is closely related to its carefully selected ingredients and unique preparation methods. The Hengxian Sliced Raw Fish has strict requirements for cooking technique. In Hengzhou, only by controlling each fillet to the thickness of 3 millimeters, weight between 8 and 10 grams and cutting the ingredients as fine as a hair and uniform in size, can you be called a true fish master.

这个过程称为“瘦身”，主要是让鱼肉更为紧实。不同的鱼制作出的鱼生的口感亦不同，青竹鱼的肉质比较细腻，红眼鲮的肉质较爽脆，鲤鱼的肉质则较为鲜美。

（2）放血

为了让片下的鱼肉薄如蝉翼、晶莹剔透，放血是绝不能少的关键工序。制作时，挑选一条活蹦乱跳的河鱼，先用刀背把鱼拍晕，接着除腮、斩尾。然后，将鱼头朝上、尾向下，挂在铁钩上放血。这时候鱼已经晕了，但还没有死。放血的时间由鱼的肉质和大小决定，为了保持鱼肉鲜美的口感，可以留一些鱼血在体内。

（3）取肉去皮

鱼血基本放尽，才开始取鱼肉和剥鱼皮。把鱼放在干燥的案板上，从鱼脊处下刀，把两侧的鱼肉割下来。在取肉的过程中，注意不能让鱼肉沾水。

去鱼皮也很讲究，一般从鱼头处轻轻切开口子，然后抓住鱼皮用力一拉，整块鱼皮就剥下来了。鱼肉去皮后，用鱼生专用纸包裹鱼肉数分钟，吸干鱼肉里渗出的水分和残血，这样鱼肉就会变得透亮。然后，片鱼红（即把血散不掉的红色的鱼肉切掉），去排骨，剩余的肉留待切片。

The best raw material to make sliced raw fish is Yujiang's delicious fish, such as wild grass carp, flower carp, dace, green bamboo fish and other river fish. The raw materials must be alive, fresh, clean and pollution-free. The caught fish cannot be used immediately to make sliced raw fish and must be cleaned for several days. In order to make the sliced fish fillets as thin as a cicada's wings and crystally clear, bloodletting is a key process that must not be skipped. To handle an alive river fish, firstly, stun the fish with the back of a knife and remove the gills and tail. Then, hang the fish on an iron hook to bloodlet it with its head up and tail down.

放血、取肉、去皮，整个过程一般要在 3—5 分钟内完成，尽可能保证鱼的鲜美滋味。

（4）切片

鱼生好吃与否，关键还在于师傅的刀工。鱼生切法分为“单飞”“双飞”“顺花”3 种。一切一断叫“单飞”，这种最为常见；两切一断叫“双飞”，即将两片鱼肉展开，如展翅欲飞的蝴蝶双翼；“顺花”切法则是为了满足经验老到的食客，横排竖排几道刀法后即成，鱼片较厚，口感更为醇厚肥美。

横县鱼生 ◆摆盘之精美，堪称艺术品

Hengxian Sliced Raw Fish ◆The exquisite presentation is worthy of being called a work of art

切片过程要注意手法的轻巧快捷，手指轻按鱼肉，手掌悬空，防止手温影响鱼肉品质。切的时候一定要直刀切，不能斜刀切，否则会影响鱼肉的纹路和口感。切的过程讲究一气呵成，看准便下刀，平均每秒切 3 刀。另外，要注意鱼生不能切得太薄，起码要有一根牙签的厚度才能保持最佳的口感和鲜味。

（5）装盘

横县鱼生还以摆盘精美和吃法独特闻名天下。在师傅的巧手之下，一片片生鱼片被制作成“年年有余”“鸳鸯戏水”“龙凤呈祥”等各类主题的花式摆盘，如同一幅精美的艺术作品，给食客带来视觉上的冲击与享受。

横县鱼生的作料多达 20 多种，完美诠释了“食不厌精，脍不厌细”的饮食遗风。青料有薄荷、鱼腥草、假蒟、柠檬叶、紫苏叶、生姜与胡萝卜等；酸料有酸柠檬、山黄皮、酸辣椒、酸藠头和酸姜等；调料有花生油、生抽与胡椒粉等。将各种作料切成如发丝般精细，摆入白瓷盘中，色彩丰富，令人食指大动。

据余富介绍，横县鱼生最早是乡下人家的“土菜”，是逢年过节或喜事宴请时的必备菜。最初，鱼生只是随意摆放，作料也就三五种，在市场化发展过程中，才开始慢慢讲究摆盘，从简单的摆圆盘，

Whether the fish is delicious or not, the key lies in the master's knife work. The cutting technique of sliced raw fish is divided into three types: "single flight", "double flight" and "smooth flower". The lightness and speed of the technique should be paid attention to during the slicing process. Use fingers to gently press the fish meat with the palm suspended in the air to prevent the hand temperature from affecting the quality of the fish. Hengxian sliced raw fish is also famous for its exquisite presentation and unique way to eat.

横县鱼生 ◆多样的作料，丰富了鱼生的口感

Hengxian Sliced Raw Fish ◆A variety of ingredients enrich the taste of sliced raw fish

发展到了摆花式，如今摆立体图案，样式越来越丰富。作料也发展到了 20 多种，主要是考虑食物寒热属性的调和及健康功效。例如，人们认为，鱼肉性凉，佐以暖胃生热的生姜、胡椒粉等，即可与之中和，而鱼生作料中的紫苏叶、鱼腥草等中草药，具有清热解毒、祛湿除燥等功效……这些药食同源的作料，使食用鱼生的层次感更丰富，同时达到一种口感与养生的平衡。

夹起一片莹白如雪的鱼生，在用花生油、酱油等调和的酱汁中迅速一捞，拌上满满的作料，一口吃下，鱼肉鲜脆爽口，毫无腥味，作料香辣酸脆，口感丰富。一品入心，再品难忘。

传承：将“横州美食”做成“中华美食”

“萧萧云叶落盘面，粟粟霜卜为缕衣”，遥想千年之前，斫鲙（切生鱼片）集精致饮食与表演艺术于一体，引得文人墨客纷纷吟诗作赋，留下无数脍炙人口的诗词佳句，食物和文化在中华大地上交织成一段灿烂的历史。

历经时代变迁及饮食习惯的变化，曾经星燎原之势的食鱼生习俗仅余下点点星火。如今，横县鱼生依然保留了孔夫子“脍不厌细”的饮食遗风。

There are more than 20 kinds of ingredients for Hengxian Sliced Raw Fish, which are divided into green and acidic materials, mainly consideration of the harmony of food properties and health effects. For example, cold fish meat is accompanied by ginger, peppers and other ingredients to warm the stomach and neutralize the cold. These ingredients with medicinal effects make sliced raw fish more multi-layered, and a balance between taste and health can be achieved. Grab a piece of fish that is as white as snow, and mix it with a full of ingredients. The fish is crispy and refreshing without any fishy smell and the ingredients are spicy and sour, which is unforgettable.

20多年来，余富一直在制作鱼生的道路上潜精研思，不仅将粤菜的摆盘艺术植入鱼生摆盘中，还不断丰富鱼生的作料，规范鱼生的制作技艺，参与起草了《横县鱼生制作技术规范》，促进行业标准化、健康化发展。

同时，余富还致力于横县鱼生制作技艺的传承和传播，通过师徒制的形式，亲传亲授，带出了一批广西鱼生餐饮品类从业者，可谓“桃李芬芳”。他还时常到职业院校给学生们授课，将“非遗”饮食带进校园。

在余富的记忆里，鱼生代表着逢年过节的喜庆味道和亲友团聚的烟火气息，他最大的目标，就是把这道“横州美食”做成“中华美食”。余富曾代表广西去参加2019东北亚（中国·延边）文化旅游美食周，其制作的横县鱼生赢得许多食客的赞誉，这让他深感自豪。下一步，他希望通过速冻鱼生及包装技术的研发，突破即做即食的时空限制，让更多人品味这道历久弥新的地方佳肴。

如今，横县鱼生作为广西的一道“非遗”美食，受到了越来越多的关注。中央电视台、广西地方媒体等纷纷探访报道，地方政府层面也大力推动，通过持续举办“横县鱼生美食节”“鱼生名刀争霸

Yu Fu has been studying and thinking on the road of making sliced raw fish for more than 20 years, not only implanting the plating art of Cantonese cuisine into that of sliced raw fish, but also constantly enriching the ingredients and standardizing the production technique of sliced raw fish. In addition, Yu Fu is actively committed to the dissemination of Hengxian Sliced Raw Fish's production technique.

横县鱼生 ◆视觉上和味觉上的盛宴

Hengxian Sliced Raw Fish ◆A feast for both the eyes and the palate

赛”等活动，促进行业发展，扩大社会影响力，将横县鱼生打造成为茉莉花之外的又一张地方名片。

“盈尺之鱼，出水欲飞”，“鱼脍之风”从2000多年前吹到现在，还将继续风行下去。

Sliced raw fish represents the festive taste of the New Year and the atmosphere of family and family reunion and friends gathering. Yu Fu's biggest hope and goal is to make this "Hengxian cuisine" into "Chinese cuisine". Nowadays, as a Guangxi intangible cultural heritage food, Hengxian Sliced Raw Fish has attracted more and more attention.

荔浦芋扣肉：芋肉相融　酥香软糯

Steamed Sliced Pork with Lipu Taro: the best soft and glutinous taro and the greasy and mellow pork

"非遗"饮食名片

◆ **荔浦芋扣肉制作技艺**

类别：传统技艺

级别：自治区级"非遗"代表性项目

批次：第八批

地域：广西桂林

"一家煮芋十里香，玉帝遥闻恨断肠"，这句民谣所说的就是荔浦芋扣肉。荔浦芋扣肉是桂北地区的传统名菜，是酒席上不可缺少的一道菜。荔浦芋扣肉是用荔浦芋与带皮的五花肉制成的。芋头松软芳香，猪肉肉质肥

Steamed Sliced Pork with Lipu Taro is a famous traditional dish in northern Guangxi, which is indispensable for banquets. The taro tastes soft and fragrant, and the pork is rich and mellow, which is suitable for all ages. In 2021, Steamed Sliced Pork with Lipu Taro Making Technique was selected into the eighth list of representative projects of Intangible Cultural Heritage at the autonomous region level in Guangxi.

扫码看视频

荔浦芋扣肉　◆甜糯香软的荔浦芋搭配肥瘦相间的五花肉，堪称“天仙配”

Steamed Sliced Pork with Lipu Taro　◆The combination of sweet, glutinous, and fragrant Lipu Taro with well-balanced fatty and lean pork belly is truly a match made in heaven

而不腻，风味独特，老少皆宜。2021 年，荔浦芋扣肉制作技艺入选广西第八批自治区级“非遗”代表性项目名录。

溯源：古时贡品 现代佳肴

荔浦芋因盛产于荔浦而闻名，故称“荔浦芋”，至今已有 300 多年的种植历史。据记载，清朝康熙四十八年（1709 年），荔浦芋由福建漳州一带传入荔浦，首先栽于县城城西关帝庙一带，并向周边辐射种植。荔浦四面环山、雨水丰富，得天独厚的自然条件使得此处栽培的芋头个头大、产量高，集香、酥、粉、糯、甜、软、鲜于一身，被誉为“芋中极品”，在清代就成为广西珍稀的“皇室贡品”。

1996 年播出的电视剧《宰相刘罗锅》中，荔浦芋让乾隆帝吃得停不下来，隔着屏幕也能让人垂涎三尺，这个地方特产由此火了一把。此后，中央电视台《旅行家》《金土地》《农经天地》等栏目对荔浦芋进行多次报道，荔浦芋名声大振。2005 年，荔浦芋获得地理标志产品保护。2008 年，荔浦芋被选为北京奥运会指定专用芋头。

优质食材成就了“荔浦芋扣肉”这道传统名菜。据传，清嘉

Lipu Taro is famous for its high yield in Lipu over 300 years, so it is called Lipu Taro. Lipu is surrounded by mountains and very rainy. This unique environmental condition makes high yield of taro here. And it became known as "the ultimate delicacy among varieties of taro" for the characteristics of fragrant, creamy, glutinous, sweet, soft and fresh. And it became a rare "royal tribute" in Guangxi during the Qing Dynasty.

庆年间，广西桂北厨师用荔浦芋与猪肉制成了荔浦芋扣肉，成为桂北一带居民婚嫁和节日席上必不可少的特色名菜。据记载，民国二十五年（1936年），蒋介石到广西巡视，桂系将领白崇禧特地用这道广西名菜招待蒋介石，蒋介石吃后赞不绝口。

荔浦当地流传着这样的民谣："一家煮芋十里香，玉帝遥闻恨断肠；五花猪肉来相伴，如来也会争品尝。巧妇细作夺天工，美食名肴天下扬。"历经数百年的传承，荔浦芋扣肉早已是荔浦及周边地区家喻户晓的传统佳肴，融于市井街舍，现于节庆宴席，登于酒店高堂，传于四海八方，蕴含着团圆的好寓意和扣住好运的好兆头，成为当地一张靓丽的饮食名片。

技艺：先炸后蒸　芋肉相融

荔浦芋扣肉是用油炸过的芋头、五花肉和桂林豆腐乳、三花酒及胡椒粉等调料一起腌制后蒸制而成的，吃起来酥香软糯，味道浓郁鲜美，是荔浦人的"心头好"和"看家菜"。

荔浦芋扣肉制作技艺自治区级代表性传承人黄学科一家五代人均是制作荔浦芋扣肉的高手，其伯父黄雄飞是四邻八村、远近闻名的红白喜事掌勺人。由于从小耳濡目染，长大后又跟着伯父学习，

High-quality ingredients have contributed to the traditional famous dish of "Steamed Sliced Pork with Lipu Taro". After hundreds of years of passing-down, Steamed Sliced Pork with Lipu Taro has become a fantastic local food card. Steamed Sliced Pork with Lipu Taro is made of fried taro, pork belly, Guilin fermented bean curd, Sanhua Liquor, pepper and other seasonings. After marinating and steaming, it tastes creamy, fragrant, soft, glutinous, mellow and delicious, and it is the favorite of Lipu people. It contains the good meaning of reunion and the good omen of luck.

荔浦芋 ◆有“芋中极品”的美誉，在清代就是广西珍稀的“皇室贡品”

Lipu Taro ◆It enjoys the reputation of being "the ultimate delicacy among varieties of taro" and was a rare "royal tribute" in Guangxi during the Qing Dynasty

黄学科逐渐掌握了这道菜的制作精髓。2015 年，因看中农村酒席的市场前景，黄学科组建了“荔浦宏门黄家·酒席”一条龙服务团队，成为当地知名的制作荔浦芋扣肉的行家里手。

黄学科介绍，荔浦芋扣肉的制作技艺几百年来基本上没什么变化，始终保持用料、工序、口味大致相同，好吃的诀窍在于选择正宗的荔浦芋。正宗的荔浦芋体形椭圆，形似纺锤，芋皮粗糙呈棕色，皮上节与节之间的距离比较短，切开后可以看到肉质上有明显的如大理石一般的紫色纹路，这是影响最终口感的关键因素。

（1）选材

荔浦芋扣肉的主要食材是猪肉和荔浦芋。猪肉要选肥瘦相间的带皮五花肉，当地人称之为“五层楼”。荔浦芋一般选单个重量在 1 千克左右的，根部大、顶部尖，并且没有发芽的，这样的芋头味道香浓，口感软糯。

（2）准备

大火起锅烧水，水开后转中火，把五花肉放入锅里煮 20 分钟。五花肉煮熟后要及时捞出来，防止肉皮散开，影响口感。用牙签或竹签在猪皮表面扎很多小洞，再抹上白醋和盐，这样油炸时猪皮会

The main ingredients of Steamed Sliced Pork with Lipu Taro are pork and Lipu Taro. The pork should be moderately fat pork belly, which is called "Wucenglou" by the local people. Authentic Lipu Taro are oval shaped, like a spindle. The taro skin is rough and brown, the distance between the nodes is relatively short, and the purple marble-like texture can be seen when cut up. Generally, choose the taro with a weight of about 1 kg, a large root, a pointed top, and not sprouting.

起泡，做出来的扣肉皮也比较蓬松。

以桂林豆腐乳、三花酒、生抽、蚝油、白糖、胡椒粉、五香粉和葱白等为原料，制作腌制五花肉的酱汁，以备后用。

（3）油炸

在锅里放大半锅油，开大火烧至六成油温，把五花肉皮朝下放进锅里油炸，这样既能使皮炸得更酥脆，又可以避免瘦肉部分炸得太柴。炸的过程中可以把油淋到露出来的肉上，让整块扣肉熟得更均匀。大概炸 7—8 分钟，直到猪皮表面出现小小的气泡，色泽呈金红色即可捞出。放入开水里浸泡 10—20 分钟，猪皮吸水而膨胀，从而去除部分油腻，使皮质松软，方便切件。

把芋头去皮后切成 7 厘米长的片状，放入七成热的油锅里，炸至金黄后捞出备用。炸过的芋头在蒸煮的过程中不会散架，还能充分吸收酱汁和猪肉的香味。

（4）蒸煮

把炸好的五花肉切成跟芋头差不多厚度的小件，放入调好的酱汁中，用手抓匀，腌制 15 分钟左右，让肉入味。再将入味的五花肉与酥香的芋头片相间着放入大碗里，注意皮要朝下，碗中淋上 2 勺清水，将其放入锅中蒸 1 个小时。蒸的过程中不要打开盖子，

Put the pork belly into the wok and cook for 20 minutes. When the meat is cooked, it should be fished out in time. Next, use toothpicks or bamboo sticks to prick some small holes on the surface of the pork skin, and smear white vinegar and salt. Place the pork, skin-side down, into the wok to fry. In the process of frying, you can scoop up oil on the exposed surface of the pork to make it cooked more evenly. Peel Lipu Taro and cut it into slices with sides 7cm long. Fry in oil until they turn golden yellow. Next, cut the fried pork belly into small pieces about the same thickness as taro. Put them into the sauce, grasp them with your hands and marinate them for about 15 minutes.

要让蒸汽均匀包裹着扣肉。全程一定要用大火蒸，否则扣肉收缩出油，会变得很油腻。

将蒸好的扣肉取出，倒扣在盘子里，放少许葱花点缀，皮色金黄、香气四溢的荔浦芋扣肉就制作完成了。吃荔浦芋扣肉时也有讲究，夹菜时要一筷夹双，代表着好事成双、大吉大利之意。吃的时候，要芋头、扣肉一起吃，才能体会到芋肉相融、绵软甘醇、细腻香滑、肥而不腻的舌尖感受。

在黄学科看来，荔浦芋扣肉这道美食最大的特色在于食材搭配，荔浦芋甜糯香软，五花肉肥瘦相宜，两者搭配起来便有了“1+1>2”的味觉效果，因此被当地人称为“天仙配”。

荔浦芋扣肉不但口感美味，而且营养丰富，含有优质蛋白、淀粉、多种维生素和无机盐等成分，具有补气益肾、健脾胃之功效，能在一定程度上提高免疫力，实属醇香味美、老少皆宜、四季可食的美食。

传承：改良技艺　创新发展

如今，作为广西十大名菜之一的荔浦芋扣肉已成为当地的标志性美食，吸引了无数食客探寻品味，陈晓卿导演亦是它的粉丝：“在

Place the pork belly skin-side down with the crisp taro slices in a large bowl. Add 2 tablespoons of water and steam for an hour. Then, carefully invert the dish onto a serving plate and finally, garnish with a sprinkle of chopped green onion. At this time, the Steamed Sliced Pork with Lipu Taro is golden and aromatic. The way of eating is also exquisite. When picking up food with chopsticks, pork and taro should be picked up at one time, which represents the meaning of good things come in pairs and good luck. When eating, it is necessary for tasting taro and pork at the same time, so as to enjoy the flavor of soft and fragrant, as well as greasy and crisp.

荔浦芋扣肉 ◆芋肉相融，肥而不腻

Steamed Sliced Pork with Lipu Taro ◆Integrated with taro, the pork is rich yet not greasy

我吃过的广西菜品里，搭配最地道的是荔浦芋扣肉……肉里有芋头香，芋头里充满‘动物性’，两种食材的味道融在一起，特别香。”

在黄学科的记忆里，小时候荔浦芋比较贵，并不常见，荔浦芋扣肉也主要在红白喜事的酒席上才能吃到。桂柳地区有着“无扣肉不成宴”的习俗，当地人有时就用红薯、马铃薯甚至饼干等代替芋头来制作扣肉。如今生活虽然越来越好，但由于荔浦芋扣肉工序烦琐耗时，平时家里也不常做，更多是逢年过节时，左邻右舍凑到一起，热热闹闹地支起油锅，摆出蒸笼，一次性蒸上几十碗，再各自带回家里吃。在这种邻里乡情中，荔浦芋扣肉成为不可缺少的一碗“硬菜”，家家相传。

如今，黄学科运营的“荔浦宏门黄家·酒席”经过多年发展以及“天网”（新媒体平台）、“地网”（传单、地推等）、“人网”（口碑）的传播，在当地已经小有名气。为了更好地适应市场需求，他在传统工艺的基础上，也根据饮食习惯的改变进行调整和改良：“以前人们选择五花肉会选偏肥的，随着健康意识的提高，现在改为选择偏瘦的五花肉，这样做出来的扣肉既不失美味，又降低了油脂。”此外，黄学科对荔浦芋扣肉的制作流程进行了标准化改进，如对原料的选择、酱汁的配比、蒸煮的时间等都进行了细致总结，以便能更

Nowadays, as one of the "top ten famous dishes in Guangxi", Steamed Sliced Pork with Lipu Taro has become a local famous delicacy, attracting countless diners to explore and taste. Huang Xueke made adjustments and improvements according to the changes of dietary habits on the basis of traditional techniques. As people's health requirements increased, he changed fatty meat to lean pork belly and reduced fat intake.

好地传授技艺，保障出品质量。

随着客户定制需求的增加，黄学科及其团队还将荔浦芋扣肉制作成真空包装的成品，使这一产品突破地域限制，成为当地逢年过节赠送亲朋好友的佳品。

在黄学科看来，荔浦芋扣肉要真正实现市场化发展，工业化生产是必然路径。近年来，他尝试过以代工的方式生产荔浦芋扣肉，但在产品如何保存、如何还原口味及如何延长保质期等方面遇到了较多问题。无论是研发还是建厂，都需要大量投入，而荔浦芋扣肉作为地域风味美食，其市场空间也有待验证和开发，实现工业化生产任重而道远。

作为"非遗"技艺传承人，黄学科经常参加各种公益活动，如"非遗"进校园、进社区等活动，宣传荔浦芋扣肉制作技艺，致力于推广这道传统美食。未来，他希望能创建工厂，实现荔浦芋扣肉成品的批量生产，让更多人品尝到这道美食。

近年来，荔浦市也大力发展荔浦芋产业，制定和完善荔浦芋产业的标准体系，形成了涵盖种植、流通、贮藏与加工等在内的完整产业链。

In addition, Huang Xueke has standardized and improved the production process of Steamed Sliced Pork with Lipu taro, such as the selection of raw materials, the proportion of sauce, cooking time and so on, in order to better impart the techniques and ensure the production quality. He also made the dish into a vacuum-packed finished product, and make it become a good gift for relatives and friends on festivals. As a master of intangible cultural heritage skills, Huang Xueke often participates in various public welfare activities to publicize the production techniques of Steamed Sliced Pork with Lipu Taro, promoting this traditional delicacy.

荔浦芋扣肉 ◆蕴含着好的寓意和兆头的地方美食
Steamed Sliced Pork with Lipu Taro ◆A local delicacy that carries good meanings and auspicious omens

目前来看，相较于柳州螺蛳粉、武鸣柠檬鸭等“非遗”美食，荔浦芋扣肉的工业化程度还比较低。随着荔浦芋产业链的发展成熟，以及预制菜风口的火热，荔浦芋扣肉产业化当大有可为。

In recent years, Lipu government has also vigorously developed Lipu Taro industry and formulated and improved the standard system, forming a complete industrial chain of planting, logistics, storage and processing. With the development of the industrial chain of Lipu Taro and the popularity of pre-made dishes, the industrialization of Steamed Sliced Pork with Lipu Taro is likely to be realized.

全州醋血鸭：醋血为魂　家乡至味

Sheldduck in Vinegar Blood, Quanzhou Style: vinegar-blood as the soul, the most authentic local flavor

"非遗"饮食名片

◆ 全州醋血鸭制作技艺

类别：传统技艺

级别：自治区级"非遗"代表性项目

批次：第三批

地域：广西桂林

"六月六，仔鸭肉，炒苦瓜，浆血醋"，这首在全州代代口耳相传的古谣，讲的便是全州古代"肴味三绝"之一的醋血鸭。逢年过节或宴请宾客时，全州人餐桌上必少不了这么一道特色美食。

"On June 6, to cook duckling meat, sauté balsam pear, and add blood vinegar", the ancient ballad talks about Shelduck in Vinegar Blood which is one of the "three unique dishes" in ancient Quanzhou. When celebrating festivals or treating guests, such a specialty is indispensable on the table of Quanzhou people. In 2010, Shelduck in Vinegar Blood, Quanzhou Style Making Technique was selected into the third list of representative projects of Intangible Cultural Heritage at the autonomous region level in Guangxi.

扫码看视频

醋血鸭一般采用盛夏上市的仔鸭为主料，杀鸭留血、注入酸醋，以嫩姜或苦瓜等为配料，在出锅前倒入醋血。成品上桌，围炉而坐，酸辣鲜香，美味难忘。2010 年，全州醋血鸭制作技艺入选广西第三批自治区级“非遗”代表性项目名录。

溯源：妙手偶得　传承千年

全州位于桂林市东北部，地处湘江上游，素有“广西北大门”之称。关于醋血鸭，目前公开资料大多认为其起源于全州县文桥镇，距今已有 1700 多年的历史。

醋血鸭　◆看似“黑暗料理”，实则味蕾享受
Shelduck in Vinegar Blood　◆Seemingly a "dark cuisine", it's actually a treat for the taste buds

据全州《蒋氏族谱》、《全州县志》、毛氏墓碑及民间故事传说，三国时任蜀国大司马的全州人蒋琬之夫人毛氏去世后，葬于今文桥镇的西北郊。毛氏后来被封谥为安阳侯一品夫人，为其守墓成为蒋氏一族的规矩。

古代全州人有个习俗，在半年节（又称“尝新节”，在农历六月初六举行）这天要杀鸭子，煮新米饭，祭拜谷神，敬天狗。在晋惠帝末年（约公元300年）的一个半年节，一个守墓人因中午交班后回家太晚，为赶时间，匆忙之中，他把妻子准备醋黄瓜用的半碗酸水当成冲鸭血的盐水，把鸭血倒进了酸水碗中。之后，他又误将醋血倒进将出锅的鸭肉中，烹煮时发现锅内鸭肉散发出一股刺鼻酸味。情急之下，守墓人只好加大火，放入花椒、紫苏叶等香料遮盖味道，不料这鸭肉越炒越香，起锅后吃起来更是回味无穷，他便喊四邻五舍来品尝，大家拍手称绝。自此，醋血鸭的美名开始在全州县内外传开。

守墓人的误打误撞，成就了醋血鸭的妙手偶得，这道美食逐渐在当地百姓中流传下来，据说在晚清时还成为一道皇家菜，名噪一时。如今，醋血鸭承载着千年历史文化，早已成为当地人记忆深处独有的家乡味道。

There is no detailed literature record about the specific time when Shelduck in Vinegar Blood, Quanzhou Style originated in. At present, according to most of the public materials, it is believed to originate in Wenqiao Township in Quanzhou County, and it has a history of more than 1,700 years. There is a legend that a local grave keeper, in order to hurry on the day of the half-year festival, mistakenly poured vinegar blood into the duck meat that will be out of the wok, so he had to put peppercorns, perilla and other spices to harmonize the taste, but unexpectedly this duck became more fragrant the more it was stir-fried, inadvertently achieving this local famous dish.

技艺：醋血灵魂　健康养生

醋血鸭呈酱色，看起来像“黑暗料理”，但只有吃过的人才能体会到那份酸辣鲜香的味蕾享受。在千年传承中，全州人匠心独具、就地取材，创出了“苦瓜炒血鸭”“芋苗炒血鸭”“豆角炒血鸭”“花生炒血鸭”“芝麻血鸭”等多个品种，其中以“苦瓜炒血鸭”最为经典和流行。但不管配菜如何创新，醋血依然是这道地方特色美食千年不变的灵魂。

全州县文桥镇的蒋福财，是蒋家醋血鸭制作技艺的第五代传人，也是全州醋血鸭制作技艺自治区级代表性传承人。蒋福财的高祖、曾祖、祖父、父亲都是烹饪好手，自小耳濡目染，蒋福财 16 岁已经能炒一手好菜，后又外出学厨、打拼，成为国家级大厨。而不管蒋福财走到哪里，传播推广醋血鸭这道家乡名菜，一直是他身体力行的一件事。

蒋福财介绍，醋血鸭的烹饪，从选材至制作过程，都有很多讲究。

（1）选材

制作醋血鸭，要以土鸭为食材，尤以文桥本地小脚、小脑袋的土麻鸭为佳。因为相较于其他品种的鸭子，全州文桥鸭的肉质更为

The slices of Shelduck in Vinegar Blood, Quanzhou Style are in a color of dark reddish brown and look like "dark cuisine". But only those who have tasted it can experience the taste buds' enjoyment of sour, spicy and savory flavors. In the time of millennium passing-down, the people of Quanzhou, with unique ingenuity and local materials, created many varieties like "sauted blood duck with balsam pear", "sauted blood duck with taro seedlings" and so on, among which "sauted blood duck with balsam pear" is the most classic and popular one. But no matter how innovative the side dishes are, Vinegar Blood is still the soul of this local specialty for thousands of years.

劲道，少有脂肪，没有腥味。鸭子一般选养了3个月左右、重1—2千克的为最佳。

（2）准备

制作酸水，其方法类似泡菜的制作方法，在坛子里加入食材和作料，并加入白酒调味，一周后汁水渗出，从坛子里取出酸水。

杀鸭之前，先准备好一碗酸水，然后杀鸭取血，把血滴入酸水之中，迅速搅拌使之混合均匀，形成醋血。这是醋血鸭的“神来之笔”，有了这些酸水，鸭血即使放两三天也不会凝成块状，还可以去腥提鲜。

把鸭肉切成小块，以便入味。同时，准备生姜、大蒜、豆豉、辣椒、五花肉、花生、苦瓜和紫苏叶等配料。

（3）炒煮

锅头烧热，五花肉爆炒出油，然后放鸭肉，再放入豆豉、生姜、大蒜等一起爆炒，接着用锅盖盖住。每隔几分钟就掀开锅盖，用铲子上下翻动鸭肉，再倒入适量清水继续小火焖煮。待鸭肉煮熟时，放入苦瓜一起炒，不仅能缓解醋血带来的腥味，还有清热解暑的功效，这样反复数次直到把鸭肉渗出的水煮干为止。

这个过程需要20—40分钟，鸭子的大小不同，所需要的时间也

Native ducks should be used to cook Shelduck in Vinegar Blood, Quanzhou Style as the raw material, especially Wenqiao ducks that have small feet and small skulls. Make sour juice in the jar. Add ingredients and condiments, then add white liquor for seasoning. After a week, the juice will be separated out. Before killing the duck, prepare a bowl of sour juice from the jar, then kill the duck and get the blood. Drop the blood into the sour juice, stir it quickly and mix it evenly to get Vinegar Blood, and put it aside for later use. This is the "stroke of genius" of Shelduck in Vinegar Blood.

有所不同，所以全州人时常对馋嘴等待的孩子说一句“心急吃不了醋血鸭”。

（4）上浆

待苦瓜片变色、变透明轻薄时，加入醋血，让鸭肉上色入味，这一道工序叫“上浆”。在烹饪醋血鸭时，上浆品相如何，直接决定了菜的品质。放入醋血的时机是关键，锅要从旺火上移开，让鸭肉降温，只有这样，调好的醋血才能均匀地裹在鸭肉上，否则温度过高，鸭血会凝结成块。

待鸭血将干时，放入花生、紫苏叶等一起搅拌均匀，一锅醋血鸭便烹制成功了。直接将醋血鸭连锅端上桌，在桌上继续用小火慢慢煨，使之更加入味。人们围桌而食，酸辣鲜香的醋血鸭，让人满口生津。

蒋福财总结说，要做好醋血鸭这道菜，有三个关键点：一是酸水要浓淡酸咸适宜；二是要把握好上浆这个过程；三是要把控好烹饪的时间和火候。这些都需要经验积淀，只要有一步不到位，味道就失之千里。

“无鸭不过秋”“处暑送鸭，无病各家”，鸭子一直是我国的传统食材，在广西更是逢年过节的必备食材。醋血鸭不仅让人百吃不

Sauté the pork until the oil comes out, put the duck meat into the wok and add the black bean sauce, ginger, garlic, etc. Then saute them together, and cover the wok with a lid. Take the lid off, stir the duck meat in every few minutes, and then add some water to simmer them. When the duck is sufficiently cooked, add balsam pear pieces and saute them together with other ingredients. When the balsam pear pieces change color and become transparent and thin, add Vinegar Blood to make the duck meat be colored and flavoured. When the duck blood is about to thicken, add peanuts, perilla and other ingredients and stir them well.

醋血鸭 ◆酸辣鲜香，尽在其中

Shelduck in Vinegar Blood ◆A blend of sour, spicy, and savory flavors, all in one dish

厌，而且营养极为丰富，具有大补虚劳、清热健脾等功效。据《本草纲目》记载，鸭肉能填骨髓、长肌肉、生津血、补五脏，酸水有开胃、健脾、通窍的作用，紫苏能解肌发表、行气宽中等，再加上苦瓜（配料多选苦瓜）“除邪热、解劳乏、清心明目、益气壮阳”的功效，故食用醋血鸭不仅健康养生，而且四季皆宜。

传承：创新古法　与时俱进

醋血鸭不仅为全州人所钟爱，还被传播到了周边县市，成为桂菜当中的一道经典菜式。改革开放以后，不少全州人在南宁、柳州、桂林、广州、深圳、武汉与长沙等地纷纷开起了醋血鸭餐馆，将这道地方特色名菜推广到了更多地方，使之为越来越多的食客所喜爱。中央电视台也曾多次到访全州，将醋血鸭带到了全国人民面前，使这道传承千年的全州至味声名日增。

如今，随着全州人走南闯北，醋血鸭也跟着全州人漂洋过海。为更好地传承醋血鸭制作技艺，蒋福财除了在餐厅言传身教地带徒弟之外，还经常受邀到各地的学校讲课。如今，他有许多徒弟在各地开店创业，可谓“桃李满天下”。此外，他还参加各类美食文化大赛，凭借醋血鸭斩获多个奖项，通过赛事交流，让更多人了解和

Shelduck in Vinegar Blood can not only make people never get tired of eating, but also provides abundant nutrition. Shelduck in Vinegar Blood not only is loved by the people of Quanzhou, but also spread to the surrounding counties and cities, becoming a classic dish in Guangxi cuisine. After the reform and opening up, many people in Quanzhou opened Quanzhou Vinegar Blood Duck Restaurants in Nanning, Liuzhou, Guilin, Guangzhou, Shenzhen, Wuhan, Changsha and other places, promoting this local specialty to more places. Nowadays, as the people of Quanzhou travel north and south, the Shelduck in Vinegar Blood also crosses the ocean with them.

喜爱这道美食。

随着人们健康意识的不断提升，醋血鸭这道菜也在与时俱进地改良。蒋福财在古法基础上创新性地推出了去皮醋血鸭的做法，这样做出来的醋血鸭油脂少、更入味，在保留鲜香酸辣口感的同时，肉质更为劲道可口。

同时，随着消费习惯的不断转变，餐饮食品化的趋势日益显现，“非遗”饮食也在求新求变。早在2014年，蒋福财便尝试将醋血鸭包装后销售。经过多年探索，在工艺标准化的基础上，他期望推出醋血鸭的预制菜包装产品和零食包装产品，让这道地方名菜突破地域限制，突破餐桌限制，拓展更大的市场。

传承古法，创新技艺，激活市场，醋血鸭这道传承千年的地方名菜，在新时代依然散发着不竭活力。它不仅是地方饮食习惯的体现，更是地方历史文化的载体。在潜移默化中，对于家乡文化的认同和家乡味道的思念，成为每一个全州人难以割舍的牵挂和羁绊。在市场化推广下，醋血鸭逐渐成为外地人食而难忘的一道佳肴，成为其了解全州的一个窗口。

Shelduck in Vinegar Blood, the famous local dish that has been passed down for thousands of years, still exudes inexhaustible vitality in the new era through the heritage of ancient methods, innovation of skills, and activation of the market. It is not only the embodiment of local eating habits, but also the carrier of local history and culture. Through the imperceptible influence of taste buds, the recognition of hometown culture and the yearning for the taste of hometown have become the concern and fetters of Quanzhou people. Under the market promotion, Shelduck in Vinegar Blood has gradually become a window to know Quanzhouan and an unforgettable delicacy for outsiders.

醋血鸭 ◆千年传承，全州至味

Shelduck in Vinegar Blood ◆Passed down for thousands of years, it is the ultimate delicacy of Quanzhou

南宁老友粉：热辣酸爽　有情有义

Nanning Laoyou Rice Noodles: spicy and sour, full of heart and soul

"非遗"饮食名片

◆ **南宁老友粉**

类别：传统技艺

级别：自治区级"非遗"代表性项目

批次：第二批

地域：广西南宁

南宁有句俗话："不吃老友粉，枉做南宁人。"南宁老友粉与柳州螺蛳粉、桂林米粉并称为广西的"三大米粉"。南宁老友粉是南宁的传统美食，其口味鲜辣、汤汁香浓，夏天吃之开胃，冬天食之驱寒，深受当地人

"If you haven't tried Laoyou Rice Noodles, you are a native of Nanning in vain". Nanning Laoyou Rice Noodles is the traditional delicacy of Nanning. Its taste is spicy and the soup is aromatic. It is appetizing to eat in summer and can dispel coldness in winter, so it is popular among the locals. In 2008, Nanning Laoyou Rice Noodles Making Technique was selected into the second list of representative projects of Intangible Cultural Heritage at the autonomous region level in Guangxi.

扫码看视频

老友猪杂粉 ◆口感丰富，层次分明

Laoyou Pork Offal Rice Noodles ◆Rich in texture and layered in flavor

喜爱。2008 年，南宁老友粉入选广西第二批自治区级“非遗”代表性项目名录。

溯源：老友常临　有情有义

在广西，几乎每座城市都有自己独特的米粉，历史悠久、浸润着文化气息的桂林米粉，新晋“网红”、火到海外的柳州螺蛳粉……南宁最具代表性的米粉，则非酸辣咸香兼备的老友粉莫属。

老友粉源于老友面。汪曾祺在《五味》中曾写道：“福建人、广西人爱吃酸笋。我和贾平凹在南宁，不爱吃招待所的饭，到外面瞎吃。平凹一进门，就叫：‘老友面！’‘老友面’者，酸笋肉丝汆汤下面也，不知道为什么叫作‘老友’。”

为何叫作“老友”？据传与一个温情的故事有关。20 世纪 30 年代，一位老翁每天都光顾周记茶馆喝茶，有几天因感冒没有去，周记老板十分挂念他，便用精制面条佐以爆香的蒜末、豆豉、辣椒和酸笋等，煮成热面条一碗，送给这位老友吃。热辣酸香的面顿时让老翁食欲大增，他发了一身大汗，感冒也好了。事后老翁感激不尽，书写“老友常临”的牌匾送给周记老板，“老友面”从此得名。后来，有人将这种老友面的做法与当地人喜食

Laoyou Rice Noodles comes from Laoyou Noodles. It is said to be related to a warm story: in the 1930s, an old man visited Zhou Ji Teahouse every day to drink tea. But for a few days he did not go because he had a cold. The boss of Zhou Ji Teahouse was very worried, so he boiled refined noodles with minced garlic, tempeh, chili, sour bamboo shoot, etc. into a bowl of hot noodles and gave it to this old friend to eat. The hot and sour noodles suddenly increased the old man's appetite, then he sweated hard and became healthy. Afterwards, the old man was grateful, so the plaque of "Old Friends Frequently Visit" was presented to Boss Zhou. So "Laoyou (Rice) Noodles" got its name.

的米粉相结合，老友粉就是这样形成的。随着时间推移，老友面（粉）渐渐扬名八桂。

老友面（粉）的故事一直口耳相传，具体时间已难以考证，但大致的版本都认为其发生在民国初期，距今也有近百年历史。而故事中浓缩在一碗面（粉）里的“老友情”，成就了一个城市有情有义的美食名片。

提到新中国成立后老友粉的发展历程，不可不提中山老友面店。20世纪50年代，南宁市饮食合作总店主营老友面，其中以位于南宁市中山路小学附近的中山饮食店最为出名。当时这家门店只煮制以湿面为主的老友面，深受顾客喜爱，由此这家门店又被称为“中山老友面店”。

老友粉是从老友面慢慢发展出来的。如今，老友面（粉）店在南宁遍地开花，催生出一批煮粉界知名老品牌，如“复记老友”“舒记老友”“天福香老友”，以及一些主打老友面（粉）的新晋品牌等。

提“老友”而知“南宁”，如今“老友味道”已成为南宁这座城市的独特记忆，传承着当地的饮食文化；“老友文化”亦成为南宁独特的城市符号，传递着包容、开放、友好的南宁精神。

Laoyou Rice Noodles has now been developed all over Nanning. A number of well-known old brands were promoted in the rice noodles industry as well as some new brands. When "Laoyou" is mentioned, Nanning is called to mind. Now "Laoyou taste" has become a unique memory of Nanning, carrying the local food culture and city style. "Laoyou Culture" has also become a unique symbol of Nanning, conveying the spirit of inclusiveness, openness and friendliness of the city.

技艺：先炒再煮　猛火快攻

不同于桂林米粉和柳州螺蛳粉拌一拌、煮一煮就能吃，传统的老友粉很讲究“镬气”（锅气），要食材新鲜、现炒现煮。经过铁锅旺火烹制出的老友粉，才有那种热辣酸爽的味道。

复记老友董事长黄宇认为，老友粉最大的特色是酸辣平衡，放料的顺序和比例、火候的把握是制作的关键。

（1）选材

酸笋、辣椒、豆豉和蒜头这四样食材被称为“老友料头”，是老友粉的标准配置。据黄宇介绍，酸笋主要选用南宁当地笋，用清水泡制、自然发酵而成；辣椒要选南宁当地小米椒，一般在每年 9—11 月收（不能在下雨天收）；豆豉古称“幽菽”，有浓郁的酱香味，是一种风味独特的作料，它隐藏在酸与辣背后，成了“老友味道”的基石。此外，老友粉在调味品上也有讲究，其所用的酱油分为调色和调香两种类型，醋要用 4° 米醋。

老友粉的配菜以肉为主，一般多用猪肉、猪杂、牛肉、牛杂等，肉汁的鲜香也是老友粉鲜味的来源之一。煮粉的汤要用提前煲好的筒骨汤。粉一般选扁粉（当地人称之为“切粉”）。

Traditional Laoyou Rice Noodles is very particular about "Huoqi" (Guoqi). The ingredients must be fresh. And after they are stir-fried and boiled over a strong fire in the iron wok, the Laoyou Rice Noodles can have that hot, spicy and sour taste. The biggest feature of Laoyou Rice Noodles is the balance between sour and hot; the order and proportion of ingredients, and the grasp of heat is the key to production. The four ingredients of sour bamboo shoots, chili pepper, black bean sauce and garlic are the standard recipe of Laoyou Rice Noodles. The side dishes of Laoyou Rice Noodles is mainly meat, generally using pork, pork offal, beef, beef offal, etc.

老友粉的基础食材 ◆酸笋、辣椒、豆豉和蒜头是成就“老友味道”的关键

The basic ingredients of Laoyou Rice Noodles ◆Sour bamboo shoots, chili peppers, black bean sauce, and garlic, which are the key to achieving the "Laoyou flavor"

（2）爆炒

老友粉之所以深受食客喜欢，与它“先炒再煮”的烹饪手法密不可分，重点在灶头功夫。

在爆炒过程中，放入食材的顺序很重要——放肉、放料、放油。具体来说，用煮粉勺先打肉，再放酸笋、蒜头、辣椒和豆豉等作料。开火热锅，然后将煮粉勺里提前打好的料扣入锅底停留 2—3 秒，通过高温干锅热炝，把老友料头的本味炝出来。放油后，肉跟作料集中在锅底，用煮粉勺推锅中的各种料来回炒，炝出老友香味。注意不要一直炒，停顿 2 秒等油热，把肉香逼出来，炒一下，停一下。

（3）煮粉

随后，在锅中放入由醋、酱油等调配而成的酱汁继续爆炒，让味道充分释放。锅中加入高汤，用旺火煮开。最后扔进一把扁粉，迅速颠锅，一碗热气腾腾的老友粉就可出锅了。

制作老友粉很讲究投料精准程度，“少一分醋则不酸，多一分酱则过咸”，一碗正宗的老友粉必须同时吃出酸、鲜、辣、咸的味道。老友粉的酸，主要来自酸笋爆炒出来的酸味，再加点酸醋，可以说，酸笋是老友粉的灵魂作料。

The reason why Laoyou Rice Noodles occupy the top spot in Nanning is inseparable from its cooking method of "first stir-frying and then boiling" which focuses on the stove. In the stir-frying process, the order of putting ingredients is important — first the meat, then the materials and last the oil. Then, add the sauce (vinegar, soy sauce) and continue stir-frying to allow the flavor to be fully released. Add the broth, using high-temperature fire to bring out the aroma of all the ingredients. Finally, throw in a handful of rice noodles into the wok, quickly toss it in the mixture, and a bowl of sour, spicy, salty and delicious Laoyou Rice Noodles can be served.

老友粉的制作过程讲究猛火快攻，一气呵成。抄锅甩勺要眼疾手快，否则肉易老，粉会断，整锅粉也不够“镬气”。一碗老友粉的制作要在1分钟内完成，门店厨师在烹饪时一般需要同时掌勺两个锅。

新鲜出炉的老友粉，汤汁醇厚，炒过的豆豉香气浓郁，酸笋的酸爽融入汤汁中，再配以爆香的猪肉、猪肝等，口感丰富，层次分明。闻之酸爽鲜香，食之开胃解馋，吃完激发出一丝汗意，犹如打通任督二脉，浑身舒畅。

新鲜出炉的老友粉 ◆闻之酸爽鲜香，食之开胃解馋

Freshly made Laoyou Rice Noodles ◆The aroma is refreshing and savory, while the taste is appetizing and satisfying

老友粉中富含碳水化合物、纤维素、蛋白质、胡萝卜素、铁、磷等多种营养成分，具有开胃解腻、祛风散寒、通窍醒脑的功效，这也是坊间流传老友粉能治感冒的原因所在。

传承：保护传统　老友南宁

老友粉，不仅仅是一道美食，更是一种情怀，传承着历史文化和老南宁的味道。

据介绍，复记老友诞生于 1918 年，当时名为“合升隆”；1982 年，更名为“复生隆”，蕴含复兴家族生意、万事兴隆之意；1989 年，又更名为“复记饮食店”；2004 年，“复记老友”商标正式注册。

复记老友粉一经推出，就受到了街坊邻里的热烈好评。在复记老友如今的掌门人黄宇的记忆里，印象最深的就是那种相互帮扶的邻里情，就像老友粉一样热情暖心。复记最早的招牌“复记老友粉”，就是当时一个擅长书法的邻居题写并赠予的，现在他们依然沿用这一招牌。

一条老巷，一间小铺，一群老友，一碗煮粉，成了南宁人记忆深处最地道的味道、最温暖的回忆。

20 世纪 90 年代后，随着市场化进程的加快，老友粉产业在蓬

Laoyou Rice Noodles is not only a delicacy but also a feeling, carrying the historical culture and taste of old Nanning. In an old alley of Nanning, there is a small shop where a group of old friends gather to enjoy steaming bowls of Laoyou Rice Noodles. This has become the most authentic flavor and the warmest memory in the hearts of Nanning people.

勃发展的同时，也面临着多元化饮食冲击、品牌鱼龙混杂、老字号后继乏人、传统技艺和口味失真等一系列问题。

从2007年开始，南宁市群众艺术馆的工作人员就对老友粉手工技艺的传承开展了详细的调查，发现曾盛极一时的老友粉店日渐减少，老友粉的制作工艺和选料日益简单化，而且一些掌勺师傅年事已高或相继去世，正统的制粉工艺面临人才断档的困境。

为了保住这一传统美食的正宗风味，2010年南宁市第一个饮食类传统制作技艺传承基地——南宁老友粉（面）传承基地在老字号“共一老友粉店”正式挂牌，通过举办技能培训班、建立“传帮带”机制等方式培养传承人，延续老友粉传统制作工艺。

美食是一个城市最直观、最具烟火气的符号。近年来，南宁市把老友粉作为“一市一品”的重点品牌进行打造，推动南宁老友粉规模化、产业化、标准化和绿色化发展。在政府的扶持推动以及从业者的不断努力下，以酸笋、蒜头、豆豉及辣椒为基调的“老友口味”得到越来越多人的认可，老友粉开始走向全国，很多新晋品牌不断涌现，网红打卡潮流兴起，这让老友粉成了南宁人出街常选、外地游客来南宁必吃的美食。

由于老友粉中酸笋混合着豆豉的独特味道，一些外地食客第一

In order to preserve the authentic flavor of the traditional food, Nanning's first heritage base of traditional food production technique — Nanning Laoyou Rice Noodles heritage Base was officially launched in the time-honored "Gongyi Laoyou Rice Noodles Store" in 2010 to cultivate successors and pass down the traditional production technique of Laoyou Rice Noodles by holding skill training courses and establishing a "teaching and helping" mechanism.

次吃的时候可能会不适应。因此，一些老友粉品牌在保持传统口味之外，也尝试推出更多适合大众口味、年轻化的新品。

而南宁人也将对“老友口味”的热爱发挥到了极致，可以说万物皆可“老友”：老友鱼、老友田螺、老友花甲、老友粉虫、老友菜心……“老友口味”也成了南宁这座城市的独特口味。

此外，为了顺应餐饮食品化的发展趋势，预包装老友粉产业近年来发展迅速。

复记老友董事长黄宇在采访中介绍，其公司已将袋装老友粉和老友功能酱推向市场。目前，其公司已在武鸣自建了工厂，拥有现代化生产线，涵盖袋装老友粉、老友小龙虾、老友鸭（鸡）杂及老

老友粉 ◆夏食开胃，冬食驱寒

Laoyou Rice Noodles ◆Refreshing in summer, warming in winter

友酱菜等系列产品。

黄宇表示，建设食品生产工厂、发展袋装食品业务与传统餐饮经营相比，无论是对她个人还是团队都提出了更大挑战，但标准化、可复制的发展路径，无疑是未来企业发展壮大的方向。他们将不局限在传统老友粉这一品类，而是以“老友风味”为主开发系列产品，“有情、有味、有老友”，服务好老友粉产业链的上、中、下游产业，立足南宁，辐射广西，走向全国，做大做强老友粉产业。

“老友，老友，吃了就算朋友”，老友粉充满温情的起源故事，以及其充满市井气息的美食体验，与南宁热情友好、开放包容的城市气质极其相符，凝结为南宁特有的城市文化，呼应着“老友南宁”的文旅品牌。而正在崛起的预包装老友粉，也将开启南宁老友粉产业发展的新篇章。

人间烟火味，最暖凡人心。约上三五老友，吃一碗热气腾腾的老友粉，聊一些柴米油盐的家常琐事，享一刻轻松闲适的时光，这大抵就是南宁这座城市最吸引人的模样。

In addition to maintaining the traditional taste, some brands are also trying to launch more new products that suit the taste of the public and younger people. Nanning people also exert their love for "Laoyou" to the extreme. It can be said that everything can be made in the "Laoyou" style: Laoyou Fish, Laoyou River Snails, Laoyou Clams, Laoyou Worm-like Rice Noodles, Laoyou Chinese Cabbage... Any ingredient, as long as it is added the "Laoyou taste", will be particularly aromatic and become the unique taste of Nanning. In addition, in line with the development trend of catering food, the prepackaged Laoyou Rice Noodles industry has developed rapidly in recent years.

柳州螺蛳粉：汤鲜笋臭　酸辣爽口

Liuzhou Luosifen: fresh soup and fermented bamboo shoots, spicy and sour, refreshing to the palate

"非遗"饮食名片

◆ 米粉制作技艺（柳州螺蛳粉制作技艺）

类别：传统技艺

级别：国家级"非遗"代表性项目

批次：第五批

地域：广西柳州

"大米小珍馐，小吃大灵魂。粉好度日月，螺小赛乾坤。"柳州螺蛳粉是以螺蛳、猪筒骨及多种香料熬制的高汤为汤底，以柳州特有的软滑筋道的米粉为主料，加上酸笋、木耳丝、炸过的花生和腐竹等辅料制作而成的。凭借"酸、辣、鲜、爽、烫"的特点，柳州螺蛳粉频频"出圈"，收获了众多粉丝，被称为"网红第一粉"。2008 年，柳州螺蛳粉手工制作技艺入选广西第二

With its "sour, spicy, fresh, stretchy, hot" characteristics, Liuzhou Luosifen has gained great fame and many fans and is known as "the most popular rice noodles on the Internet". In 2018, "Liuzhou Luosifen" won the geographical indication trademark. In 2021, Liuzhou Luosifen Making Technique was successfully selected as the fifth list of representative projects of the national Intangible Cultural Heritage.

扫码看视频

批“非遗”代表性项目名录。2018 年，“柳州螺蛳粉”成功注册成为地理标志证明商标。2021 年，米粉制作技艺（柳州螺蛳粉制作技艺）成功入选第五批国家级“非遗”代表性项目名录。

螺蛳粉　◆螺蛳、米粉一相逢，便胜却人间无数

Luosifen　◆When the river snails meet rice noodles, it surpasses countless delights of the world

渊源：街头市井的创造　民间智慧的结晶

关于螺蛳粉的起源，坊间流传着多个版本，最久远的可追溯到唐代文学家柳宗元喜食螺蛳粉的传说。但如今大多认为，螺蛳粉起源于 20 世纪 70—80 年代的广西柳州。其是由路边的摊贩将熬螺蛳的汤和米粉混合在一起后偶然做成的，因受到当地人喜爱而迅速发展起来。

广西人民出版社出版的《柳州螺蛳粉》一书中记载：关于螺蛳粉的起源，江湖传说无数，传说中最靠谱的故事是在 20 世纪 70—80 年代，柳州的夜生活逐渐恢复，夜市也逐步繁荣，柳州人嗜吃螺蛳和米粉，不少食客在吃米粉的时候有意无意地要求在自己的米粉里加入鲜辣螺蛳汤。久而久之，第一个聪明的摊主开始尝试用螺蛳汤煮米粉，没想到食客们吃后反应很好，夜市的摊主们有样学样，慢慢形成了现在螺蛳粉的雏形。

虽说螺蛳粉是近几十年来才兴起的，但要说螺蛳粉完全是新的美食也并不准确。因为不管是其中的“螺蛳情结”还是“米粉情结”，都极具历史渊源和柳州当地特色。

“一味螺蛳千般趣，美味佳酿均不及”，嗍螺一直是柳州人的爱好。据考古发现，柳州地区食螺历史可以追溯到 2 万多年前。柳州

Nowadays, it is mostly believed that Liuzhou Luosifen originated in Liuzhou, Guangxi in the 70s and 80s of the 20th century. At that time, it was accidentally made by a street vendor who mixed the soup of boiled river snails with rice noodles. It was loved by the local people and then developed rapidly. Although Liuzhou Luosifen has emerged in recent decades, it is not accurate to say that it is completely new. Because both "love for river snails" and "love for rice noodles" have historical origins and show local characteristics of Liuzhou.

白莲洞洞穴科学博物馆出土了大量螺蛳化石，大多数螺蛳的尾部存在被人工敲掉的痕迹。学者推测，遗址中与螺壳同时期出现的敲砸器，可能就是敲砸螺壳尾部的工具。也就是说，白莲洞人懂得采集、加工和食用螺蛳。穿过2万多年的光阴，柳州人加工螺蛳时先敲掉尾部的习惯依然保留着，煎、炒、蒸、煮，调香入味，十八般厨艺下培育出来的“螺蛳情结”与极具历史渊源的“米粉情结”巧妙结合，最终产生了螺蛳粉这道美食。

螺蛳粉的另一大主角——米粉同样历史悠久。广西是稻作文化发源地之一，衍生出了多彩的米食文化。广西的米粉样式很多。早在唐朝，就有御赐“交州米饼”的记录，那时的交州包括现在的广西，所谓“米饼”即线条状的米粉食品。《徐霞客游记·粤西游日记二》中也有关于柳州人“竟不买米，俱市粉饼食”的记载，这表明明朝柳州已有喜食米粉之俗。及至现代，米粉依然是广西人深爱的主食之一。

时代的因缘际会下，螺蛳与米粉的相逢，便成就了螺蛳粉这道地方美食。虽然螺蛳粉最先出自谁手无法溯源，但其作为街头市井的创造、民间智慧的结晶是毋庸置疑的。可以说，螺蛳粉见证了当地食用螺蛳与米粉的历史，承载着一方百姓的味觉记忆，

Eating river snails has always been a special hobby of Liuzhou people. According to archaeological discoveries, the history of river snail-eating in Liuzhou can be traced back to more than 20,000 years ago. Rice noodles also have a long history. Guangxi is one of the birthplaces of colorful rice planting culture, in which various kinds of rice noodles are the most popular. Liuzhou Luosifen status as the creation of street vendors and the wisdom of common people is beyond doubt. It can be said that Liuzhou Luosifen witnesses the history of local people eating river snails and rice noodles, and it is a special carrier of local collective memory.

是地方性集体记忆的特殊载体。

技艺：工序繁复 螺蛳汤是精髓

经过不断改良，螺蛳粉制作技艺逐渐成熟。制作螺蛳粉，需以螺蛳为主料熬制粉汤，烫上米粉，再辅以酸笋等各种辅料。制作过程注重食材选取和火候掌握，讲究细节把控。

螺蛳粉制作工序繁复，螺蛳需吐泥，笋要发酵，米粉经浸米、磨浆等传统工序制成。熬制螺蛳汤所用的辅料配比精细，经炝炒、熬煮、冲油再熬煮而成，大火、文火应把控有度。

在柳州螺蛳粉制作技艺自治区级代表性传承人张泰华看来，一碗螺蛳粉吃的是味、品的是道，单是汤头就熬足 18 个小时，正是这种匠心，才成就了螺蛳粉的“网红”之道。

（1）选材及准备

螺蛳粉的食材全部取自柳州当地。螺蛳主要选用柳州当地的石螺，以拇指大小为宜，肉质肥美鲜甜，这种螺在河流、溪水及池塘里都很常见。捞起螺蛳后先把它们放在清水中静置一两日，让它们吐泥。为了加快这一过程，还会在水中泡上一把菜刀或其他铁器，这样煮出来的螺蛳不会有浓重的泥腥味。

The river snails are local river snails, which are suitable for thumb size. This kind of river snail meat is plump, fresh and sweet. Rice for milling is made by mixing aged brown rice with newly oily sticky rice, which is made into "dried rice noodles" through multiple processes. As for the ingredients, sour bamboo shoots are indispensable. It is mainly made by selecting sweet and plump bamboo shoots in spring, picking, peeling and other fine procedures, and then adding spring water and fermenting them in a large earthen jar for about half a month. Fermented sour bamboo shoots will emit a faint odor, which is the source of the unique "stink" of Luosifen.

制粉的米选用隔年糙米和新出油粘米混合，用泉水浸泡后磨浆，经过多道工序制成“干榨粉”，粉煮前浸泡6—12小时，仍保留其米味香浓，口感滑韧。

辅料方面，酸笋是必备的。选取肉质饱满的大头甜笋，经过采摘、剥皮等精细的处理后，加入山泉水，用大瓦缸腌制半个月左右，酸笋就腌制好了。发酵后的酸笋会散发出淡淡的腐臭味，这正是螺蛳粉特有“臭味”的来源。

其他的辅料选配任搭，一般有炸过的花生、腐竹，以及酸豆角、木耳丝、黄花菜和萝卜干等，讲究香脆的口感。鸭肠、鸭脚和猪脚等则是锦上添花之物，丰俭由人。

米粉的选材　◆选用柳州当地生产的“干榨粉”

Selection of rice noodles　◆Using "dry pressed rice noodles" locally produced in Liuzhou

螺蛳粉的辅料 ◆丰俭由人，惊艳味蕾
Toppings for Luosifen ◆A variety of options to suit any budget, delighting the taste buds

鲜香的螺蛳汤 ◆螺蛳粉的精髓
Fresh and fragrant River Snail Soup ◆The essence of Luosifen

（2）熬汤

螺蛳汤是螺蛳粉的精髓，每家所用食材及其配比各有不同，属于独家秘方，但制作流程基本相同。先将螺蛳剪去尾尖，焯水冲洗后用铁锅干炒出味，与已用油煸炒出香的十几种香料猛火爆炒 3 分钟，使螺蛳吸足“前香”；加水没过螺蛳，文火收汁，使螺蛳吸足“后香”，让螺蛳的鲜味更浓。

在此基础上，加入水与猪筒骨，以及酸笋、茴香、紫苏、沙姜、八角、桂皮与草果等一起熬煮，先用大火熬煮 1 小时，然后转文火熬煮 6 小时。这样熬煮出来的高汤既有石螺的鲜味、猪筒骨的精华，又有香料的浓香。最后在汤底中加入辣椒油，用文火调制而成螺蛳汤，此时的螺蛳汤鲜、辣、烫，且带有回甘，具有清而不淡、麻而不燥、辣而不火、香而不腻的独特风味。

（3）煮粉

清水烧开，放入米粉，煮好后捞出装碗，加入酸笋、酸豆角、木耳丝、炸过的花生与腐竹等辅料，再舀上几勺红彤彤的螺蛳汤，一碗热气腾腾的螺蛳粉即可品尝。米粉软滑筋道，浓汤酸辣鲜爽，咬一口还没被汤浸透的腐竹片，又香又脆，尤其是备受争议的酸笋，带着浓郁悠长的“臭味”，入口后又清爽酸脆，瞬间上

River Snail Soup is the essence of Luosifen. Each family uses different ingredients and proportions, but the production process is basically the same. First, cut off the tail tips of the river snails, wash them with water, and then fry them in an iron pan. After that, stir-fry the river snails and other spices that have been added to the oil for 3 minutes, so that the river snails can absorb the flavor of spices. Then add water to boil the river snails with pork bones, sour bamboo shoots, fennel, perilla, sand ginger, star anise, cinnamon, tsaoko and other flavourings together. Finally, pour chili oil into the broth base and simmer it into River Snail Soup.

色香味俱全的螺蛳粉 ◆酸辣均衡，鲜香开胃
Delicious and fragrant Luosifen ◆Balanced sourness and spiciness, fresh and appetizing

头，一尝难忘，由此流传着吃螺蛳粉只有 0 次和无数次的“真香定律”。

螺蛳粉虽然看着像重口味的“黑暗料理”，实则是高蛋白质、低脂肪的天然食品。螺蛳粉中的米粉是主食的一种，现代研究表明，米粉中含有 75% 左右的碳水化合物、7%—8% 的蛋白质、1.3%—1.8% 的脂肪以及丰富的 B 族维生素；辅料中的腐竹含有多种矿物

When the water boils, add rice noodles. After cooking, fish out into the bowl, add sour bamboo shoots, sour beans, agaric shreds, fried yuba, fried peanuts and other ingredients, and then scoop a few spoons of River Snail Soup into, a bowl of Luosifen can be tasted. This kind of rice noodles is soft and smooth, the thick soup is spicy and sour. Although Luosifen looks like a "dark cuisine", compared with junk foods, it is a natural food with high protein and low fat.

酸爽滑韧的螺蛳粉 ◆嗍上一口，爱上一生

The sour, refreshing, smooth, and resilient Luosifen ◆One bite, a lifetime of love

质，植物蛋白较高，可补充钙质，是豆制品中的佼佼者；酸笋含有丰富的纤维素，能促进肠蠕动；花生性平，根据《本草纲目》的记载，花生能悦脾和胃、润肺化痰、滋养补气、清咽止痒。

一碗螺蛳粉，了却百般愁。螺蛳粉以其浓郁的汤底和独特的口感独步天下，从一城一域的地标美食，一跃成为火爆“出圈”的美食。

Rice noodles is a staple food. Modern research shows that it contains about 75% carbohydrate, 7%-8% protein and 1.3%-1.8% fat, and it is rich in B vitamins. The yuba in the ingredients contains a variety of minerals and high vegetable protein, which can supplement calcium and is a leader in soybean products. Sour bamboo shoots are rich in cellulose, which can promote intestinal peristalsis. Peanuts are mild and have a record of "benefit the viscera and dissolve phlegm".

传承：当代柳州的独家记忆

在广西的“三大米粉”中，虽然螺蛳粉的历史最短，却以异军突起之势快速崛起，如今已成为风靡世界的“非遗”美食。回溯螺蛳粉的发展及传承路径，不难发现，政府的强力推动及工业化思维在其中起到了至关重要的作用。

自兴起于柳州夜市后，螺蛳粉便以鲜明特色快速发展。20 世纪 80 年代中期，在不同门店的竞争中，螺蛳粉的辅料不断丰富，口味逐步优化，形成了“酸、辣、鲜、爽、烫”的稳定口感。通过家族传承和师徒传承的方式，螺蛳粉不断发展壮大，涌现了一批知名的螺蛳粉餐饮品牌。

到了 20 世纪 90 年代初，螺蛳粉已经遍布柳州的大街小巷。《鱼峰文史》记载：“鱼峰辖区内现经营螺蛳粉的不下一百摊，几乎大部分街巷都有……大多摊点是日夜兼营。”

螺蛳粉餐饮业虽呈雨后春笋般蓬勃发展之状，但一直囿于广西区内，初次“出圈”则始于 2011 年。当年，柳州市政府推出“螺蛳粉进京”项目，这是螺蛳粉走向全国的第一步。2012 年，开创了美食纪录片新纪元的《舌尖上的中国》开播，第一集就介绍了柳州的酸笋和螺蛳粉：“有一种叫螺蛳粉的小吃中，酸笋是最重要的辅

In the mid-1980s, Luosifen formed a stable taste of "sour, spicy, fresh, stretchy, hot". A number of well-known Luosifen catering brands have emerged. By the early 1990s, Luosifen spread all over the streets of Liuzhou. In 2011, Liuzhou proposed the "Luosifen into Beijing" project, which was the first step for Luosifen to go nationwide.

料。”“惊鸿一瞥”的螺蛳粉开始被更多人所知，其在全国的知名度得到快速提升。

螺蛳粉餐饮业真正的转机在于2014年。随着米粉制作工艺的提升以及物理杀菌与真空包装等食品生产技术的引入，当年10月，柳州商人严振华获得柳州市原食品药品监督管理局颁发的全国首张袋装螺蛳粉生产许可证，自此螺蛳粉正式开启产业化进程。

螺蛳粉进入“袋装速食”时代后，搭上了互联网电商的快车，地方美食逐渐走向更广阔的天地。2018年，螺蛳粉登上了某电商平台发布的“民间美食地图”第一名。2020年疫情防控期间，螺蛳粉更是成为火遍全国的现象级产品。

随着螺蛳粉的持续火热，螺蛳粉经营者在竞争中不断推陈出新，细分出原汤螺蛳粉、干捞螺蛳粉、炒螺蛳粉和凉拌螺蛳粉等多个产品；在辅料搭配上独出心裁，推出三鲜螺蛳粉、花甲螺蛳粉等不同花样；甚至跨界创新，螺蛳粉口味的饼干、月饼、青团与粽子等层出不穷。对南宁人来说，万物皆可“老友”；对柳州人来说，则是一切皆可“螺蛳”。

政府层面也不断加大扶持力度，如指导成立柳州市螺蛳粉行业协会，编写柳州螺蛳粉地方标准，加大螺蛳粉餐饮店全国布局的力

In 2014, with the introduction and improvement of food production technologies including rice noodles production technology, physical sterilization, and vacuum packaging, Luosifen officially started the industrialization process. With the continuous hot sales of Luosifen, the operators are constantly bringing forth new ideas in the competition.

度，构建螺蛳粉全产业链，建设螺蛳粉产业园、螺蛳粉特色小镇、螺蛳粉电商产业园等产业集聚区等，从各方面完善配套措施，为螺蛳粉产业发展“保驾护航”。

在此基础上，通过“总部 + 基地”“龙头企业 + 专业合作社 + 农户”等模式，引导周边县区大力种植竹笋、木耳、豆角等，养殖螺蛳。柳州市通过打造螺蛳粉原材料基地，带动农民增收，使螺蛳粉产业成为当地产业发展、乡村振兴的新引擎。

同时，柳州市推动“螺蛳粉 + 旅游”深度融合。2021 年，柳州市推出全国首条螺蛳粉旅游线路，打造出一条“观螺蛳粉文化馆—游螺蛳粉生产线—品螺蛳粉”的特色工业旅游线路，充分挖掘螺蛳粉的历史渊源及文化内涵，展示了螺蛳粉产业发展和文化科技创新的最新成果。

螺蛳粉产业的爆发式发展，使得人才需求更为迫切。2020 年 5 月，全国首家螺蛳粉产业学院正式成立。不同于笼统的厨师职业教育和社会学徒式传承，该学院采用“职业教育 + 产业”的新模式，培养了一批高素质、极具创新力的螺蛳粉产业技术技能复合应用型人才。

一碗螺蛳粉延伸出产业链，成为“非遗”饮食项目创造性转化

The government has also continuously increased support, guided the establishment of Luosifen Industry Association, compiled local standards for Luosifen, expanded the layout of Luosifen restaurants throughout the country, and improved supporting measures from all aspects to "escort" the development of the Luosifen industry.

和创新性发展的代表性案例。2022 年 3 月，国家乡村振兴局发布了《圆梦新时代　开启新征程——2021 年全国巩固脱贫成果和推进乡村振兴典型案例掠影》，其中《打造螺蛳粉全产业链，夯实产业基础——广西柳州市螺蛳粉产业振兴新探索》成功入选全国 14 个典型案例掠影之一。

“非遗”保护需要政府主导，“非遗”传承更需要群众支持。为了更好地传承与发扬柳州螺蛳粉制作技艺，柳州市相关部门创新性地推出“非遗”研学品牌活动，打造螺蛳粉研学传习所、举办螺蛳粉制作技艺大赛等，让螺蛳粉文化走进校园、融于生活。

螺蛳粉不仅是柳州人的“心头好”，也因其独特口味征服了全国食客的胃。习近平总书记在考察柳州时是这样称赞螺蛳粉的：“甜、酸、苦、辣、咸，各种味道都有了。不分南北，大家都喜欢吃。”

某短视频平台发布的《2022 非遗数据报告》显示，在全国各地最受欢迎的国家级“非遗”美食排名中，螺蛳粉位居第一。

“不食螺蛳粉，枉为柳州人。”无论去过多少地方，见过多大的世界，尝过多少美食，螺蛳粉总在一头牵绊着柳州人。而螺蛳粉产业的发展壮大，正是敢想敢干、敢为人先的“柳州精神”的一个生

The protection and promotion of intangible cultural heritage needs the guidance of government and the support of people. The relevant departments of Liuzhou have innovatively launched the intangible cultural heritage research brand activities, and built the Luosifen research and study institute, so that the Luosifen culture can enter the campus and integrate into life. Luosifen not only is the favorite of Liuzhou people, but also conquers the stomachs of diners all over the country because of its unique taste.

螺蛳粉 ◆层次丰富，回味无穷

Luosifen ◆Multi-layered, leaving an unforgettable aftertaste

动缩影。

如今，螺蛳粉文化的影响已延伸至海外，螺蛳粉已成为柳州这座工业老城又一个闻名遐迩的城市符号，折射出当代柳州在探索中的创新路径。螺蛳粉，已然乘风起，未来更可期。

The development and growth of the Luosifen industry is a vivid epitome of the "Liuzhou spirit" that thinks big and dares to be the first. Today, Luosifen has become another well-known symbol of Liuzhou, an old industrial city, reflecting the innovative path of contemporary Liuzhou in exploration of development.

桂林米粉：卤水鲜香　米粉滑韧

Guilin Rice Noodles: rich and fragrant sauce, smooth and chewy noodles

"非遗"饮食名片

◆ **米粉制作技艺（桂林米粉制作技艺）**

类别：传统技艺

级别：国家级"非遗"代表性项目

批次：第五批

地域：广西桂林

桂林山水甲天下，桂林米粉誉全球。桂林米粉历史悠久，有着独特的文化底蕴和制作工艺。桂林米粉是用桂林所产大米及漓江水为基本原材料制作而成的，以独特配方熬制的卤水为灵魂，以卤菜粉流传最广，深受当

Guilin's landscape is the best in the world, and Guilin Rice Noodles is famous all over the world. With a long history, Guilin Rice Noodles has its unique culture and production techniques. It takes the rice produced in Guilin and the water of Lijiang River as the basic ingredients, accompanied with the boiled marinade according to the private formula of historical restaurants in the city. Among various kinds of Guilin Rice Noodles, the rice noodles with braised food are the most famous and loved by local people.

扫码看视频

桂林米粉　◆桂林甲天下的不止山水，还有米粉
Guilin Rice Noodles　◆Not only is Guilin renowned for its magnificent landscapes, but also for its delicious Rice Noodles

地人喜爱。2021 年，米粉制作技艺（桂林米粉制作技艺）入选第五批国家级“非遗”代表性项目名录。

溯源：半壁民国一碗粉

关于桂林米粉的起源，至今没有确切的考证，坊间普遍认为起源于 2000 多年前，与秦始皇派兵修灵渠有关。中国当代著名国学家、文化学者肖健将其总结为：“‘六王毕，四海一’，百越平，米粉现。”

相传，秦始皇吞并六国、统一中原之后，立即挥师 50 万大军南下征战百越。为迅速平定岭南，他下令在桂林开凿灵渠。然而，南下征战的秦军多为北方人，他们吃不惯南方的饮食。秦军将领便想出办法，“南粮北做”，把大米磨成浆，加工成像北方面条一样的米粉条，这种米粉条受到将士们的欢迎，桂林米粉的雏形由此形成。

另外，由于长时间在南方丛林中行军，水土不服，许多秦军将士中暑、感冒，患上了疟疾等疾病。军中医生为了给将士们治病，用传统中草药与当地人提供的香料等煎熬制成防疫药汤，让将士们服用。由于战事紧张，将士们经常是米粉、药汤合在一起吃，久而久之，药汤就成为桂林米粉卤水的雏形。经过历代改进加工，其慢

In 2021, Guilin Rice Noodles Making Technique was selected into the fifth list of representative projects of the National Intangible Cultural Heritage. As for the origin of Guilin Rice Noodles, there is no conclusive textual evidence to track down. It is generally believed that Guilin Rice Noodles originated over 2,000 years ago and was related to the fact that Qin Shihuang, the first emperor of the Qin Dynasty, had built the Lingqu Canal.

慢演变为如今别具风味的桂林米粉卤水。

对于“秦代修灵渠秦军发明桂林米粉”的说法，学术界一直存在争议。支持者从各种细节去印证这种说法，如：桂林米粉文化艺术馆中陈列的最原始的米粉压榨机，上面的“离合器”完完全全就是秦时农具“耒”的样子；过去卖担担米粉的小贩穿着的木板鞋，正是秦时履的一种。种种证据，不一而足。

反对者则从各种文献考证中加以反驳。例如，原载于《社科文萃——2016年桂林市社科学术年会论文集》中的《论“桂林米粉”品牌形成的历史及原因》一文中，作者依据国学大师钱锺书《管锥编》的考据，结合相关史料记载及文物界的考古成果认为：米粉应该最早出现在东汉时期的江西一带，后流传到南方各省。

桂林米粉到底起源于何时仍处于争议之中，但桂林米粉兴盛于清朝末年以及民国年间则是不争的事实。

清朝时，桂林轩荣斋的炒粉、会仙斋的卤菜粉、易荣斋的汤粉，各有各的特色，吸引了无数回头客。三斋之间不是以拆对方的台而后快，而是互相勉励，各出奇招：你卖汤粉，我就卖卤菜粉；你卖卤菜粉，我就卖炒粉……三斋之间既公平竞争，又不断翻新花样，所以那时流行着一句行话：“米粉拌匀谓之和。”“三斋打擂”之事，

The issue that "Guilin Rice Noodles was invented by the Qin army when building the Lingqu Canal in the Qin Dynasty" is always controversial in the academic circle. Supporters confirm this statement from various details. Opponents refute it from various textual research and they believed that the earliest rice noodles appeared in Jiangxi during the Eastern Han dynasty. Later, it spread to the southern part of China.

让桂林米粉声名远播，至今仍为人津津乐道。

到了民国年间，尤其是抗日战争时期，随着新桂系的崛起，以及大量文化名人聚集桂林，桂林米粉更是名声大噪。在那段历史的浮沉变幻中，桂林米粉一次次出现在时代舞台上：抗战文化城里文人的马肉米粉经、李宗仁的“米粉交际”、白崇禧白公馆里的米粉故事……从市井记忆到政坛风云，桂林米粉承载了一个时代的记忆，正可谓“半壁民国一碗粉”。

这时期，最广为人知的，莫过于马肉米粉和担担米粉。当时装马肉米粉的碗只有茶杯那么大，每个碗中只有一根米粉在里面（所以桂林有“吃米粉找不到头”之说），鲜美的马骨汤配香脆的腊马肉片，再加芫荽、花生、芝麻油，那香味，直冲肺腑。民间素有“不吃马肉米粉，不知天下美味”的谚语。出生于桂林的当代作家白先勇先生就曾在文章里追忆：“哈盛强的马肉米粉最出名，我一口气可以吃五六碟，吃了回来，抹抹嘴，受用得很。”

马肉米粉虽然美味，但量少价高，非寻常百姓所能常吃的，担担米粉则更受欢迎。担担米粉（也称“担子米粉”）是因将桂林米粉装在担子里，挑着它走街串巷叫卖而得名的。担子的一头装着火炉、卤水锅，另一头装着米粉、配料等。这种充满着市井烟火气息的担

The origin of Guilin Rice Noodles is still controversial, but it is the fact that Guilin Rice Noodles flourished in the late Qing Dynasty and the Republic of China. In the Qing Dynasty, Xuanrongzhai's fried Rice Noodles, Huixianzhai's Rice Noodles with braised food and Yirongzhai's Rice Noodles with soup had their own characteristics and attracted numerous customers. The story of the competition among the three restaurants became a folklore. In the Republic of China, especially during the War o f Resistance Against Japanese Aggression, with the rise of the new Guangxi clique and the gathering of a large number of cultural celebrities in Guilin, Guilin Rice Noodles became more famous.

子米粉，在抗日战争那个动乱时期，滋养了不少因躲避战乱而迁居桂林的文人。丰子恺先生曾写文章称赞这类担子，说其兼具艺术性与实用性，堪称“微型厨房”。

桂林米粉制作技艺自治区级代表性传承人龙付孙最早就是跟着师傅，挑着担子，沿街卖粉。龙付孙介绍说：“担担米粉可以称为桂林卤菜米粉的祖师，它起源于清末民初，盛行于抗日战争时期，从 20 世纪 50 年代开始衰落，到 20 世纪 80 年代后期基本消失在街头巷尾。”

如今，担担米粉早已走进了店内，硬件环境不断升级；传统的“老四样桂林米粉”（卤菜粉、马肉粉、原汤粉、牛腩粉），也已演化为以卤菜粉为代表、涵盖数十个细品的大品类，并以其独特口味，享誉全国乃至全球。

技艺：特色在于鲜香　精华集于卤水

桂林米粉做工考究，制作一碗桂林米粉需要经过鲜粉生产、卤水熬制、辅料制作、调配出品四大步骤，共有百余道工序，其中鲜粉生产与卤水熬制最为关键。据食家言，桂林米粉之妙，在于卤水之香之鲜，粉线之软滑而又有“筋力”。

During this period, the most widely known was Horse Meat Rice Noodles and Dan Dan Rice Noodles. The delicious horse bone soup was accompanied by crispy slices of smoked horse meat, coriander, peanuts and sesame oil, in which the fragrance went straight to the heart. The production of Horse Meat Rice Noodles was few and the price was high, which was uncommon for ordinary people to eat, so Dan Dan Rice Noodles was more popular. "Dan Dan" means carrying pole with one end loading a stove, a pot of brine, the other end loading rice noodles, ingredients, and so on.

龙付孙从 1983 年开始，师从当时被誉为“米粉世家”传人的李桃年，学习桂林米粉及卤水制作技艺，到如今已经 40 余年。在他看来，与其他米粉相比，桂林米粉的特色在于鲜香，精华集于卤水。

（1）鲜粉生产

广西自古便盛产水稻，桂林更是广西的水稻大市，素有“桂北粮仓”之美誉。明代著名地理学家徐霞客游历桂林时就曾写下“月色当空，见平畴绕麓，稻畔溢水，致甚幽旷”的文字，描述了稻田连片的场景。

传统的桂林米粉都是手工制作的，从稻米变成米粉，前后需要经过清洗、浸泡、发酵、磨浆、熟化、成型、蒸粉、冷却等多道工序。

先将桂林产的早籼米、晚籼米、陈米按 3：2：5 的比例用漓江

桂林米粉 ◆沿用古法制作的米粉既软滑又有“筋力”

Guilin Rice Noodles ◆Produced by traditional methods, these rice noodles are soft and smooth, resilient and chewy

水浸泡 3 天，然后将米洗干净并磨成浆，用纱布包好压干水分，再将压干的米浆搓成粉团，下锅煮至三分熟七分生，然后放入石臼中捶打均匀后，放入器具中压制成米粉，煮熟后再过一遍冷水即可备用。沿用传统古法制作的桂林米粉具有滑而不黏、韧而不硬、绵而不糜的特质。

不同于螺蛳粉的干粉，桂林米粉的粉主要是鲜湿的圆头米粉或者切粉。鲜米粉只有 4—6 个小时的保鲜时间，所以桂林的米粉厂要凌晨开工，以保证早上各个米粉店能够按时收到米粉。

（2）卤水熬制

一碗米粉好吃与否，最关键的还是卤水是否正宗。卤水的雏形是以祛病为目的的药汤，它经过漫长发展和技艺演化，逐步添加数十种药用食材和辅料，最终形成了如今的卤水。

正宗的桂林米粉的卤水，都是用多种草药和香料熬制而成的，无添加剂、无色素、无香精。但具体用料和配比，每家都有自己的秘方。

卤水的制作过程可以说是调和百味，火候和时间的把控是关键。龙付孙所传承的卤水制作技艺属于“发酵型卤水”制作技艺。用这种技艺制作卤水时，采用罗汉果、沙姜、花椒、小茴香、白芷、

Guilin Rice Noodles is exquisite in production techniques; it includes more than 100 steps such as the production of fresh rice noodles, the boiling of marinade, the production of ingredients, and the final cooking, among which the first two steps are the most important. Traditional Guilin Rice Noodles is hand-made through cleaning, soaking, fermenting, grinding, ripening, shaping, steaming, cooling and other processes. Unlike the dry rice noodles of Luosifen, Guilin Rice Noodles is mainly round rice noodles or cutting rice noodles, fresh and wet.

甘草、香叶、桂皮、丁香、陈皮和草果等 20 多种草药和香料，按照传统比例配比，经过清洗、炒制后，放入三花酒，加入清水，再加入牛骨、牛肉或是猪骨、猪肉等大火猛煮，煮开后再文火熬制，将肉卤熟后捞起。

经过 9 个小时熬制的卤水要经过两次恒温静置，龙付孙将这个过程称为“发酵”。第一次发酵 48 小时，过滤掉草药和香料渣滓后，在不同温度下再继续发酵 7 天，这个发酵过程是很多卤水制作所没有的。

如此制出的卤水清香淡雅、温和纯正，看上去颜色浓厚，拌入米粉却不着痕迹。

（3）辅料制作

如果说卤水是桂林米粉的灵魂，锅烧（炸酥的五花肉）是精华，那么卤菜就是伴侣了。以其中的锅烧为例，卤好的五花肉第一次油炸时皮朝上，稍微冷了后，再用 200℃的高温皮朝下进行第二次油炸至起泡，到了吃的时候，还要回锅进行第三次油炸。

为什么要炸三次？龙付孙介绍，第一次是固定形状，第二次是让肉皮起泡，第三次则是为了保持肉的脆度，因为锅烧超过 4 个小时就会回潮。

The taste of Guilin Rice Noodles depends on the marinade. The embryonic form of marinade is a decoction for the purpose of healing. After a long period of development and evolution, dozens of medicinal ingredients and trimmings are gradually added, which finally forms today's marinade. It can be said that the production process of marinade needs to harmonize various tastes, in which the control of heat and time is the key.

热制卤水的草药和香料　◆卤水鲜香的关键

The herbs and spices used in brewing the marinade　◆The key to the delicious and fresh aroma of marinade

（4）调配出品

等到吃桂林米粉的时候，需要用30℃的温度加热卤水，使卤水的香味更好地挥发出来。米粉则要用85℃的水烫一下（桂林人称之为“冒粉”）。烫好的米粉滤干后放入碗中，配以锅烧、卤牛肉片和叉烧等，加卤水、花生油、酥黄豆，再根据个人喜好加入辣椒、蒜蓉、葱花、酸豆角与酸笋等，搅拌均匀，一碗色香味俱全的卤菜粉就做好了。

桂林米粉在吃法上讲究“一粉两吃”——先干捞，让卤水的香味充分释放，让每一根米粉都裹匀了卤水，吃到一半或者吃完后再加入用新鲜骨头熬制的清汤。地道的桂林米粉，其粉条圆细而柔韧，卤水芳香鲜美，肉菜香松爽口，“根根有头不见头，碗碗有汤不见汤”，吃完唇齿留香，念念不忘。

桂林米粉不仅美味，而且富有营养。米粉富含蛋白质、铁元素与碳水化合物等成分，补血益气、健脾暖胃，适于脾气虚弱的人群食用。卤水则用多种草药和香料熬制而成，既可调味提香，又具理气化湿、解表祛风的保健作用，具有药食同源的功效。

The production of braised food is also very fine. For example, Guoshao, also known as fried streaky pork, should be fried with the skin facing up for the first time. After slightly cooling down, it should be fried at a high temperature of 200℃ with the skin facing down until puff up. When it is time to eat, it should be fried for the third time. Rice noodles should be scalded with 85℃ of water. Drain the hot rice noodles and put them into a bowl with Guoshao, braised beef slices, barbecued pork, etc., and then add marinade, peanut oil and crisp soybeans. Next, you can add chili, garlic, scallion, sour beans, sour bamboo shoots and other ingredients according to your personal preferences.

桂林米粉　◆若说锅烧是精华，那么卤菜就是伴侣

Guilin Rice Noodles　◆If Guoshao is the essence, then the braised food are the companions

传承：桂林人一辈子的乡愁

桂林山水宜人，用漓江水滋养出的优质稻米制作而成的米粉也透着清雅闲适的气息，渗透到了桂林人的日常生活。

“幸福生活哪里有？二两米粉三两酒。”在20世纪80年代，随着担担米粉消失在街头巷尾，桂林米粉店如雨后春笋般涌现，其环境、配套设施不断升级。每天早上去店里吃一碗桂林米粉，“老板，多给点黄豆，再帮打三两三花酒”，这样的叫喊声在当时的米粉店此起彼伏，成了很多老桂林人的记忆。

进入21世纪，人们的工作和生活节奏不断加快，很多习惯已经改变，但嗍粉的习惯一如往昔。去个相熟的小店，或坐或蹲，吃一碗粉，喝几口汤，开始元气满满的一天。中餐、晚餐不知道吃什么的时候，嗍粉永远是不会出错的选择。

桂林米粉不仅是桂林人爱得深沉的一道日常食物，喂饱了当地人的胃，还是在外游子割不断的乡愁，填满了桂林人的心。

在这里不得不提白先勇先生，他虽然只在桂林生活了7年，但是一直眷恋着故土，尤其酷爱桂林米粉。从他的小说《花桥荣记》《玉卿嫂》等中都能看到他的桂林米粉情结。在《少小离家老大回——我的寻根记》中，白先勇先生更是直白地写道：“吃的东西，

In the 1980s, with the disappearance of Dan Dan Rice Noodles, Guilin Rice Noodles began to enter the restaurant, and the dining environment was constantly upgraded. Every morning going to the shop to eat a bowl of Guilin Rice Noodles and drink some Guilin Sanhua Liquor became the memory of many elder Guilin people. After entering the 21st century, the pace of work and life continues to, and many customs have changed, but the one of "eating Rice Noodles" is the same as before.

桂林别的倒也罢了，米粉可是一绝。因为桂林水质好，榨洗出来的米粉，又细滑又柔韧，很有嚼头……我回到桂林，三餐都到处去找米粉吃，一吃三四碗，那是乡愁引起原始性的饥渴，填不饱的……外面的人都称赞云南的'过桥米线'，那是说外行话，大概他们都没尝过正宗桂林米粉。”

不尽的惆怅、浓浓的乡愁，穿过时光沉淀，镌刻进了一碗桂林米粉中。2015 年，美国有线电视新闻网（CNN）评选出 14 道最令中国人魂牵梦绕的“思乡菜”，桂林米粉名列榜单第 4 位。

正是这份味蕾上的羁绊，让桂林米粉慢慢走向全球。20 世纪 90 年代后，在北京、上海、广州和深圳等城市，桂林米粉店遍地开花，甚至还开到了国外。近些年，随着米粉保鲜保质的关键性技术瓶颈被突破，速食装桂林米粉产品开始出现在国内大型连锁超市和便利店。

龙付孙是改革开放之后桂林较早出来自主创业的个体户，其经营“明桂米粉”这个品牌已有 30 多年了。作为“非遗”技艺传承人，龙付孙近年来一方面通过招收徒弟、进校讲学等方式，传播桂林米粉文化，传承古法制作技艺；另一方面不断强化产品研发，突破技术限制，积极发展预包装产品，将桂林米粉推向更大市场。

Mr. Bai Xianyong has only lived in Guilin for seven years, but he has always been sentimentally attached to Guilin, especially Guilin Rice Noodles. His Guilin Rice Noodles complex can be seen from his novels, such as *My Rice Noodle Shop* and *Madam Yu Ching*. Guilin Rice Noodles ranked fourth in CNN's list of 14 "homesick dishes" that haunt Chinese people in 2015. It is the strong sense of geographical belonging brought by the taste buds that makes Guilin Rice Noodles slowly go global.

在他看来，技术上的创新都是提高效率的手段，核心的东西是坚决不能改变的，那就是真材实料、保持最初的味道。他最大的愿望，就是将桂林米粉推向全世界，“一辈子只做一碗粉，让世界尝尝桂林的味道”。

从“一碗乡愁”到国家级“非遗”，桂林米粉承载着历史变迁，承载着文化往事，承载着古法技艺，成为桂林除山水之外的又一张城市名片。

嗍一碗米粉，忆往昔峥嵘岁月，看今朝世间繁华。

As the successor of the intangible cultural heritage of Guilin Rice Noodles Making Technique, in recent years, Long Fusun has spread Guilin Rice Noodles culture and imparted the ancient production techniques by recruiting apprentices and lecturing in schools. On the other hand, he continues to strengthen product research and development to break through technical restrictions, and actively develop pre-packaged products of Guilin Rice Noodles for a larger market.

南宁生榨米粉：软糯香滑　开胃健脾

Nanning Squeezed Rice Noodles: soft, glutinous, fragrant, and smooth, appetizing and good for the spleen

"非遗"饮食名片

◆ **南宁生榨米粉制作技艺**

类别：传统技艺
级别：自治区级"非遗"代表性项目
批次：第六批
地域：广西南宁

碾米舂团榨成丝，鲜酸软糯酵香醇。生榨米粉又称"榨粉""生榨粉"，是将大米发酵后，经过多个工序加工煮制而成的一种米制品。其以软、滑、香的口感，以及发酵产生的与众不同的"酸馊味"而闻名，吃上一口，软糯酸爽，让人回味无穷。2016年，南宁生榨米粉制作技艺入选广西第六批自治区级"非遗"代表性项目名录。

Squeezed Rice Noodles is a kind of fermented rice product, which is made by fermenting rice and cooking it through multiple processes. It is famous for its soft, smooth and fragrant taste, as well as the distinctive "sour taste" given by lactic acid bacteria after fermentation. In 2016, Nanning Squeezed Rice Noodles Making Technique was selected into the sixth list of representative projects of Intangible Cultural Heritage at the autonomous region level in Guangxi.

扫码看视频

溯源：民俗节庆的必备美食

生榨米粉是广西壮族地区的一道特色美食，在南宁老城区、邕宁、武鸣、马山、上林及河池大化等地最为常见。公开报道资料显示，生榨米粉源于明末清初时期，每到收获时节和民俗节日，它就成为当地百姓必备的一道美食。生榨米粉是广西稻作文化、节庆文化、祭祀文化的典型代表，至今已有数百年的历史。

民国广西《同正县志》记载："四月初八日为浴佛节，各家多做米粉，以为过节。米粉有切粉、线粉之分。以暖水泡米至数日，发酵、细磨、滤干、舂和，用器榨成粉条于沸汤内，制似桂林米粉而稍异，和原汤，配鱼肉、酱油、香料等熟食，味带酸，城厢人长为之。"同正县，明清时称"永康州"，隶属南宁府。

据说，生榨米粉最初是蒲庙"花婆节"上的一种庙会小吃，随着到蒲庙逛庙会的人越来越多，蒲庙生榨米粉的名声逐渐壮大。如今每年农历三月十一至十二，在蒲庙"花婆节"上，展示生榨米粉的传统制作技艺已经成为固定节目。

在南宁，久负盛名的还有水街生榨米粉，它深受食客喜爱。清末民初，南宁邕江沿岸水街的饮食业呈现一派繁荣兴旺的景象，形成了水街独具特色的传统饮食文化。经过百年时光沉淀，生榨米粉

Squeezed Rice Noodles is a specialty of Zhuang ethnic minority in Guangxi. According to public reports, Squeezed Rice Noodles originated in the late Ming and early Qing Dynasties. During every harvest season and folk festival, Squeezed Rice Noodles becomes an indispensable delicacy. It is a typical representative of rice planting culture, festival culture and sacrificial culture in Guangxi with a history of hundreds of years.

生榨米粉 ◆发酵型米制品，因具有独特的“酸馊味”而闻名

Squeezed Rice Noodles ◆Fermented rice product is renowned for its unique "sour taste"

如今已成为水街传统美食的典型代表。

南宁生榨米粉历史悠久，底蕴深厚，是广西民族饮食文化的珍贵遗产。其具有选料精细、制作考究、久煮不烂、鲜香爽滑、开胃健脾、老少皆宜等特点。

技艺：自然发酵　现榨现煮

生榨米粉在壮语里被称为“粉馊”，主要是因大米经发酵散发出一种自然的“酸馊味”而得名。这种“酸馊味”则来源于其特殊的制作工艺。

南宁生榨米粉制作技艺自治区级代表性传承人黄天玲是地道的南宁人，她自小就跟从母亲学习制作生榨米粉，是家族第五代传承人。20 世纪 80 年代，黄天玲在南宁水街创办了“古源香天天生榨米粉”品牌，至今已有 40 多年。

据黄天玲介绍，制作生榨米粉要用传统技艺，需经过泡米、磨浆、压干、蒸粉、舂粉和压榨六道工序才能完成，每一道工序都决定着米粉的品质。

（1）泡米

选用存放 1 年左右的陈米，加入清水浸泡，让其自然发酵，夏

Squeezed Rice Noodles is called "rice noodles with sour taste" in Zhuang dialect, mainly because the rice is fermented and emanates a natural sour taste. The production of Squeezed Rice Noodles requires the traditional techniques, including soaking rice, grinding, draining, steaming, kneading and squeezing, each of which determines the quality of rice noodles.

天放上 1—2 天即可，冬天天气冷，发酵时间需要更久，有时还要淋上热水，提供合适的发酵温度。

发酵这一过程，对发酵时间和温度的把控要求较高，时间太短则“酸馊味”不足，太长则易霉变。发酵好的大米有米香味，用手揉捏米粒时呈细碎粒状。

（2）磨浆

将浸泡后的大米，洗净后放进磨里，研磨成乳白色米浆。据黄天玲介绍，最早的时候磨浆完全是靠人工手磨，特别辛苦，现在经过技术创新，已经改为电动磨浆。

（3）压干

将米浆装入白布袋后，加重压把米浆中的水分沥至半干。以前多用石头压，一般要等一个晚上，中途还要翻搅两次，如今有了专用的压干设备，节省了很多时间和精力。

（4）蒸粉

将沥干的米粉团掰碎，再重新揉捏成拳头大小的粉团，置于沸水中或煮或蒸 20 分钟左右。蒸煮到粉团表皮 1—2 厘米深熟透，如柚子皮般即可捞起。

It is necessary to select aged rice stored for about one year, soak it in clean water and let it ferment naturally. In summer, it only needs to be kept for 1 to 2 days. In winter, when the weather is cold, the fermentation will be longer, and sometimes hot water should be poured to provide the appropriate fermentation temperature. Wash the soaked rice and put it into the mill and grind it into milky white rice paste. After the rice paste is put into a white cloth bag, it is necessary to add heavy pressure to drain the water until it is almost dry. Drain the rice noodle dough, break it into pieces, and then knead it into small dough balls about the size of fists. Put them into boiling water to cook or steam for about 20 minutes.

（5）舂粉

将煮熟的粉团放入石臼，舂捣成糊状，加入适量水，揉到完全融合不黏手，即成为榨粉时用的粉团。如今用机器来研磨，只需制粉师傅把握好粉与水的比例以及研磨时间，即可制出细腻均匀的粉团。用这样的粉团榨出的米粉才会柔韧，粗细均匀，表面平滑。

（6）压榨

将粉团放入传统的手工压榨器内，压榨出粉条并入锅煮 1 分多钟，煮熟后用长筷与漏勺捞出即可食用。如今，很多食客不喜黏糊糊的口感，会在捞出米粉后再过一次热水，让其更加爽滑，更有弹性。

出锅的生榨米粉粉质细腻，口感筋道，在碗里淋上一大勺碎肉末、卤汁，配上“老三件”：葱花、生菜碎段及切成细丝状的紫苏叶，一碗生榨米粉便大功告成。食客再根据个人口味加上花生、酸笋、酸豆角与柠檬酱等，使之口感更为丰富，让人吃后回味无穷。

生榨米粉现榨现煮，较为新鲜。由于加工过程中发酵所产生的乳酸菌，其有一种与众不同的“酸馊味”。据南宁食品营养专家的研究解读，生榨米粉发酵所产生的乳酸菌，通过利用和分解米里面的营养物质，如蛋白质、脂肪与淀粉类等物质，产生了一些类似

It is necessary to put the cooked rice ball into a stone mortar, pound them into paste, then add appropriate amount of water to knead them until they are completely fused and no longer sticky. Then put the rice ball into the traditional manual squeezer, squeeze out the rice noodles and boil it for one minute or two. After cooking, pour a tablespoon of minced meat and marinade in the bowl, accompanied by chopped green onion, lettuce and perilla, and now a bowl of Squeezed Rice Noodles is finished. Because of the lactic acid bacteria produced by fermentation during processing, it has a distinctive sour taste.

生榨米粉　◆现榨现煮，故称“生榨”

Squeezed Rice Noodles　◆It is squeezed and cooked freshly on the spot, thus it is called "Squeezed Rice Noodles"

氨基酸产物及短链脂肪酸、有机酸等物质，赋予生榨米粉独特的风味。

生榨米粉富含碳水化合物、脂肪、蛋白质等人体每天必需的营养物质，其天然产生的乳酸菌，有利于肠内微生物分布正常化，帮助维持肠道环境的健康，促进身体新陈代谢，具有增进食欲、刺激肠胃、促进消化吸收和解暑祛湿等功效，老少咸宜，深受喜爱。

传承：在改良中创新发展

相较于南宁老友粉、柳州螺蛳粉与桂林米粉，生榨米粉的知名度并不是很高，更多还是为老南宁人所熟悉和热爱。

究其原因，主要在于很多外地人接受不了那种“酸馊味”。此外，生榨米粉要现榨现煮，工艺流程繁复，经验技术要求高，相较于其他米粉，更难实现便捷化和标准化。且生榨米粉店长期经营于小镇或老街，在品牌度、宣传度等方面都不足，犹如“养在深闺人未识”的小家碧玉。

对此，黄天玲有着切身感受和清醒认知。在 40 多年的从业生涯中，黄天玲在保留传统技艺流程和风味的基础上，不断进行着创新和改良。在口味方面，她将传统发酵改良为温控操作，缩短了发

Compared with Nanning Laoyou Rice Noodles, Liuzhou Luosifen, and Guilin Rice Noodles, Squeezed Rice Noodles is well known to and loved by the older generation of Nanning people, but not very famous. The reason is that many outsiders and new urban residents can not accept the "sour taste". In addition, Squeezed Rice Noodles needs to be squeezed and cooked freshly on the spot, which is very complicated, and requires experienced and skilled chefs, so it is more difficult to achieve convenience and standardization than other rice noodles. Moreover, Squeezed Rice Noodles shops are operated in small towns or some old streets all the time, which are insufficient in brand and publicity.

生榨米粉的配料 ◆多样的配料，让口感更丰富
The toppings of Squeezed Rice Noodles ◆Various ingredients that enrich the taste experience

酵时间，减轻了米粉中的“馊味”，使酸味更加醇香，又增加了一些调味配菜，推出五色原生态生榨米粉、生料生榨米粉等创新产品，使其更适应南来北往的客人的口味，更符合年轻人的需求。在经营方面，古源香天天生榨米粉走上了品牌连锁经营的道路，门店环境和管理经营水平得到了很大提升。在生产方面，古源香天天生榨米粉通过提高机械化程度，促进生榨米粉的标准化生产，并通过“中央工厂＋门店配送”的方式，保证了门店出品质量的稳定性。

此外，黄天玲还经常到学校举办生榨米粉制作技艺培训班，与南宁市壮志路小学共建生榨米粉“非遗”传承基地等，教授生榨米粉制作技艺。其位于水街的经营生榨米粉的老店，也成为“非遗”传承基地。下一步，黄天玲希望建一个工作室，把生榨米粉制作技艺教给更多年轻人，吸引他们成为生榨米粉产业发展的生力军。

在媒体传播环境日渐多元化的今天，作为地方美食的代表，生榨米粉有了更多的“出圈”机会，尤其在明星推荐、食客探店等效应的加持下，生榨米粉获得了越来越多的关注，受到了越来越多食客的喜爱。2018 年，广州市发布《米其林指南》，这是继上海之后第二个发布美食指南榜单的内地城市。将南宁蒲庙正宗的生榨米粉带到广州的“靓姐南宁蒲庙生榨米粉”店，连续多年登上米其林的推荐榜单。

近年来，政府层面也加大了对南宁生榨米粉产业的支持力度。自 2017 年开始，南宁市邕宁区举办了多届“生榨米粉文化旅游美食节”，通过“美食 + 旅游 + 文化”的形式，将生榨米粉打造成为邕宁乃至南宁本土文化的新名片。

In today's increasingly diversified media environment, as a representative of local cuisine, Nanning Squeezed Rice Noodles has more opportunities to become famous. It is becoming increasingly popular among diners. In recent years, the government has also offered more support for the Nanning Squeezed Rice Noodle industry. Through the layout of "Food, Tourism, and Culture", Squeezed Rice Noodles has become a new business card of Yongning district and even Nanning city's local culture.

六堡茶："红、浓、陈、醇" 千年茶味

Liupao Tea: "red, strong, aged, mellow" , a thousand years of tea flavor

"非遗"饮食名片

◆ **黑茶制作技艺（六堡茶制作技艺）**

类别：中国传统制茶技艺及其相关习俗

级别：联合国教科文组织人类"非遗"代表作

时间：2022 年

地域：广西梧州

"遥远苍梧野，莽莽六堡山。山水美且秀，名茶出此间。"六堡茶，属黑茶类，主要分布在广西梧州市苍梧县六堡镇。

六堡茶历史悠久，传承千年。清嘉庆年间，其以

扫码看视频

Liupao Tea, which belongs to dark tea, has a long history. During the reign of Emperor Jiaqing in the Qing Dynasty, it was selected as one of China's 24 famous teas for its four unique characteristics of "red, strong, aged, mellow" flavors and its distinctive betel nut aroma. In modern times, it has become renowned overseas as a "tea for overseas Chinese".

“红、浓、陈、醇”四绝及独特的槟榔香味入选中国 24 个名茶之列，成为朝廷贡品；近代以来，其作为“侨销茶”闻名海外。2014 年，黑茶制作技艺（六堡茶制作技艺）入选第四批国家级“非遗”代表性项目名录；2022 年，入选联合国教科文组织人类“非遗”代表作名录。

溯源：盛于明清　侨销海外

六堡茶因地得名。据《广西通志稿》记载，六堡茶，产苍梧。苍梧茶叶出产之盛，以多贤乡之六堡及五堡为最。六堡茶原产于广西梧州市苍梧县六堡镇，产区坐落在北回归线带北侧，群峦耸立，云雾缭绕，年平均气温在 21℃左右，年降雨量充足。特殊的地理环境，造就了六堡茶与众不同的特色和底蕴。

苍梧产茶历史悠久，《桐君采药录》中有这样的记载：“南方有瓜芦木，亦似茗，至苦涩，取为屑茶饮，亦可通夜不眠。煮盐人但资此饮，而交、广最重，客来先设，乃加以香芼辈。”旧时的“交、广”就是现在的广东、广西和越南北部一带，交州州府设在苍梧郡广信县，即如今的梧州市。

明朝时，苍梧分为 11 个乡，下设堡、闸、洲和甲。六堡隶属多

In 2014, Liupao Tea Making Technique was selected into the fourth list of representative projects of the National Intangible Cultural Heritage. In 2022, it was included in the UNESCO Representative List of Human Intangible Cultural Heritage. Liupao Tea is named after the place. It originated in Liupao Town, Cangwu County, Wuzhou City, Guangxi. It is the special geographical environment here that create the unique characteristics of Liupao Tea.

六堡茶　◆汤色红润透亮，素以“红、浓、陈、醇”著称

Liupao Tea　◆The tea soup is ruddy and transparent, and it is well-known for its "red, strong, mellow and aging" characteristics

贤乡管辖。

及至清朝，六堡镇已经是南方著名的茶叶产区。据史料记载，在清康熙三十二年（1693年）的一张苍梧地图上，六堡镇所属区域就被明确地标为茶亭（当时茶商和茶农的交易场所）。清康熙三十六年（1697年）版《苍梧县志》记载："茶产多贤乡六堡，味厚，隔宿不变……茶色香味俱佳。"到了清嘉庆年间，六堡茶入选中国24个名茶之列，成为朝廷贡品。

到了清朝末年，福建、广西、广东一带的大量华人劳工"下南洋"从事锡矿开采工作。许多华人劳工水土不服，常有中暑、风湿、痢疾等疾病发作。六堡茶因其具有祛湿解毒、调理肠胃等功效迅速在华人劳工中流行，并被当成药来喝，素有"保命茶"之称。据说，当年矿场如不提供六堡茶则难以招到工人。

这样的需求带动了六堡茶向东南亚地区出口，这便是"茶船古道"的开始：从苍梧县六堡镇的河口码头驶入，直达广东省肇庆，再由东安江驶入西江，途经珠江，驶至广州十三行，远销东南亚地区。正是靠着这条航道，六堡茶得以走出深山，远销海外，成为当时著名的"侨销茶"。

当时的六堡镇茶园漫山遍野，家家户户都种茶制茶，可以说是

In the Ming Dynasty, Liupao is under the jurisdiction of Duoxian township. By the Qing Dynasty, Liupao Town had already been a famous tea-producing area in the south, and Liupao Tea was selected as one of the 24 famous teas in China. In the last years of the Qing Dynasty, a large number of Chinese laborers "went to Southeast Asia" to be engaged in tin mining. Because of Liupao Tea's effect of dehumidification, detoxification and so on, it quickly became popular among the workers and was used as a medicine which was known as "life-saving tea".

盛况空前。每到采收季节，来自广州、佛山的茶商便在六堡镇合口街开设分号、茶庄，茶号按级别定好价，直接开秤收茶，高峰期大小茶商有二三十家。有关资料显示，1935 年广西出口的茶叶达到 1128.95 吨，其中很大一部分是销往东南亚地区。

然而，战争改变了这一切。二战爆发后，六堡茶产业元气大伤。

中华人民共和国成立后，百废待兴。1953 年，梧州茶厂成立，肩负起“打开外销市场，换回六堡茶的历史声誉”的使命，首创“双蒸双压”和“冷水渥堆发酵”技术，实现了六堡茶从长期手工作坊式生产向工业化大生产的转变。此时，东南亚锡矿产业开发高潮再起，六堡茶的出口外销迎来了短暂的复兴。

1958 年开始，国内外政治及经济形势复杂多变，六堡茶产业遭受巨大冲击。及至 20 世纪 90 年代，六堡茶产业几乎湮没无闻。

“青山缭绕疑无路，忽见千帆隐映来。”进入 21 世纪，六堡茶市场迎来了新的复兴。千年六堡茶，历经沧桑起伏，如今再次扬帆起航谱新章。

However, the outbreak of World War II dealt a heavy blow to the Liupao Tea industry. In 1953, Wuzhou Tea Factory was established, shouldering the mission of "opening up the export market and retrieving the historical reputation of Liupao Tea". The export of Liupao Tea ushered in a temporary revival. However, soon after, the political and economic situation at home and abroad was complex and changeable. The Liupao Tea industry has once again been cut off. By the 1990s, the Liupao Tea industry was almost unknown. Entering the 21st century, the market for Liupao Tea has a new revival.

技艺：繁复的工序　时间的艺术

六堡茶色泽褐黑光润，汤色红润透亮，素以“红、浓、陈、醇”四绝著称。

六堡茶传统制作工艺，是相对于20世纪50—60年代出现的“双蒸双压”和“冷水渥堆发酵”的现代工艺而言的。传统制茶工艺一般包括采青、炒青、揉捻、堆闷、复揉和烘干六道工序。以六堡镇当地的原生群体种茶树所产的优质茶叶为原料、沿用传统制茶工艺制作而成的六堡茶，又称“古法六堡茶”或“农家茶”。

六堡茶制作技艺国家级代表性传承人、苍梧县六堡镇黑石山茶厂技术总监韦洁群介绍，制茶是一个循序渐进的过程，选择茶青的眼力、揉茶的力道、炒茶的时间和温度，都会影响茶叶的形状、汤色及口感。

（1）采青

六堡茶茶叶的采摘标准是一芽一叶、一芽二叶、一芽三叶，茶叶采摘必须非常小心，以免芽叶破损。韦洁群介绍，手工采茶要求提手采，保持芽叶完整、新鲜、匀净，不能捋采、抓采、掐采。采摘后的茶叶一般均匀放在簸箕上摊晾一晚，以减少茶叶的水分。

Tea making is a gradual process, and the selection of tea leaves, the strength of tea kneading, and the time and temperature of stir-frying will all affect the shape, soup color and taste of tea. The picking standard of Liupao Tea leaves is one bud with one leaf, one bud with two leaves, and one bud with three leaves. After picking, the tea leaves are generally evenly spread on the dustpan for a night. The spread-out and air-dried Liupao Tea is placed in a large iron wok heated to 200-250℃ for frying. The frying process should be done in one smooth action, without overcooking or letting the leaves to turn yellow.

采青 ◆茶叶采摘讲究芽叶完整、新鲜、匀净

Tea picking ◆The art of tea picking emphasizes the integrity, freshness, and evenness of the buds and leaves

（2）炒青

将摊晾后的六堡茶放入200—250℃的大铁锅中进行翻炒，叫“炒青”。其目的是通过高温破坏和钝化鲜叶中氧化酶的活性，抑制鲜叶中茶多酚等的酶促氧化，蒸发鲜叶的部分水分，使茶叶变软，同时去除青臭味，促进良好香气的形成。

炒青要一气呵成，不能炒焦，也不能让茶叶变黄。炒青时一般先闷炒，再扬炒，老叶多闷少扬，嫩叶多扬少闷，动作是先慢后快，逐步提高锅温。整个过程要通过听、闻、看来把控，全靠制茶人的经验积累。

韦洁群介绍：“将茶叶放进锅里，听到噼里啪啦的声音，就证明温度和火候比较适宜。入锅后要用手将茶青抖起来，把热气散出去，

炒青　◆一般先闷炒，再扬炒

Stir-frying　◆Generally, the process begins with stewing and then tossing

避免里面的茶青被闷黄。如此反复十几分钟后，看到叶色转为暗绿色，略有黏性，叶质变得柔软，拿起一个茶叶梗，反复折都不断，并能闻到茶香，就代表炒青完毕，可以出锅了。”

韦洁群说：“在炒青过程中，一定要掌握好火候。如果温度不足，茶叶干燥后滋味就会不新鲜，制成的茶汤会较浑浊；如果温度过高，茶叶易焦，后期茶叶的内含物质转化也受影响。”

（3）揉捻

将炒青后的茶叶放在簸箕内，趁热手工揉捻，以利成条，揉出茶汁，卷紧条索，再松压轻揉，回收茶汁。注意要用掌心的力度，将茶叶团向同一方向揉捻，先轻揉后重捻，手法轻重的时机把握直接影响成茶的外观和品质，十分考验制茶师傅的手艺。待茶叶放凉后，再放入大约60℃的锅里回温揉捻。反复多次，直到茶青有了固定的形状。

韦洁群说：“茶叶没有温度时不利于成形，而且容易越揉越碎，所以要通过不断回温，来保持揉捻过程中的温度。”

揉捻的目的是将茶叶揉成形，同时使叶肉细胞组织受挤压而破坏，茶汁流出附着于叶面，为渥堆过程发生物理、化学变化创造物质条件。

Place the fried leaves in a bamboo dustpan, knead them by hands while they are warm to facilitate the formation of strips. Knead out the tea juice, roll the strips tightly, and then loosen the pressure and gently knead to recycle the tea juice. When the tea leaves are cool down, put them in a wok at about 60℃ and knead them again. Repeat the action in several times until the tea leaves have a fixed shape. Then place the tea leaves in bamboo baskets or on bamboo mats for pile fermentation.

（4）堆闷

堆闷是形成六堡茶独特品质的关键性工序，即将揉捻好的茶叶放入竹箩内或堆放在竹席上进行发酵。环境的温度和茶汁的湿度会促进茶叶的内含物质转化，使叶底转色，汤色加深呈黄红色，并减轻涩味。韦洁群说：“由于整个堆闷过程不加水，制成的茶味道浓郁，汤感厚重。”

堆闷时间一般为 10—15 小时。堆闷厚度视气温高低、湿度大小和叶质老嫩而定，原则为：嫩叶薄堆，老叶厚堆；高温高湿薄堆，低温低湿厚堆。通过堆闷促使质变均匀，直至叶色变为深黄带褐色，茶坯出现黏汁，发出特有的醇香为止。

（5）复揉

经堆闷后的茶坯，有部分水分散失，条索回松，干湿不匀，需复揉一次，使茶汁互相浸润，干湿一致，条索卷紧，以利干燥。复揉前最好烘热一下，使茶坯热化回软，以利成条。复揉方法是轻压轻揉，揉到有些刺手、不能再揉为止。

复揉的功夫在慢不在快。经过反反复复揉捻，茶坯条索更为细紧，这也正是手工制作的六堡茶更为耐泡的原因。

The piling time is generally 10-15 hours. While tea leaves are piling, water is dispersed in some leaves, some strips are loose and the wetness are uneven, so re-kneading is needed to make the tea juice infiltrate all leaves. The drying of Liupao Tea is to roast the leaves on a stove which has seven fire holes over a open fire made by pine wood. Drying is carried out in two times. At this point, the ancient production process of Liupao Tea is completed.

（6）烘干

六堡茶的烘干是在七星灶上采用松柴明火烘焙，切忌以晒代烘。烘干分两次进行：第一次打毛火，烘焙温度为60—70℃，要求“高温、薄摊、快干”，使茶叶受热均匀，干燥一致，烘至六七成干后，再进行摊晾；第二次打足火，烘焙温度为50℃以下，要求“低温、

堆闷　◆形成六堡茶独特品质的关键性工序
Pile fermentation　◆A crucial step in forming the unique quality of Liupao Tea

厚摊、慢烘”，干燥程度以手折茶梗即断、叶片一捏即碎、握茶有声响为度。

至此，六堡茶的传统制作过程就算完成了。采用传统工艺制作的六堡茶在本质上属于生茶。新茶阶段的茶汤以浅黄、杏黄为主，略带苦涩，回甘好，香气层次丰富，口感偏鲜爽。要使茶叶达到相对醇和的滋味和口感，还需存放转化，由时间来为其注入灵魂。其“越陈越香”的品质，被茶友称为“时间的艺术”。

在坊间，六堡茶又称为“爷孙茶”，意思是爷爷做的茶要留给孙子喝，制好的六堡茶一般是要搁置几年以后才喝的，于是茶就代代延续下来。据说，陈年六堡茶更具有药效，民间常用陈年六堡茶加冬蜜来治疗痢疾和腹泻，所以有“三年药，五年宝，十年丹”之说。明代李时珍所著的《本草纲目》中就有相关记载：“六堡茶有清凉解热、暖胃提神、健脾助消化之功效，可治中暑、感冒、屙呕肚痛，驱除油腻，尤以陈茶为佳。”

20世纪80年代，六堡茶就曾作为保健品风靡日本。其降脂、解腻及减肥功效经日本药业界检测证实，以“健美瘦身茶”“油解茶”名称在日本各地的药店出售，引发了不少日本专家学者对六堡茶的研究和考察。日本茶界知名人士丰茗会董事长松下智还撰写了

In recent years, Liupao Tea has become more and more popular among tea lovers, not only because of its mellow and refreshing taste, but also because of its miraculous health-keeping and medicinal properties. In the 1980s, Liupao Tea was popular in Japan as a health care product. Its lipid-lowering, anti-greasy and slimming effects have been confirmed by the Japanese pharmaceutical industry.

六堡茶的存放　◆六堡茶越陈越香，堪称“时间的艺术”

Storage of Liupao Tea　◆Liupao Tea becomes more fragrant as it ages, truly a "masterpiece of time"

一本关于六堡茶的书籍，详细介绍六堡茶的特点和功效，并倡导人们吃过油脂多的料理后饮用六堡茶。六堡茶易入口、不易变质、不妨碍睡眠等特点，使之成为日本人饭后常饮的健康饮品。

清末名士程远道曾诗赞六堡茶的保健功效：“六堡名茶满山岗，止疴去腻有专长，请君泡碗今宵喝，明目犹留齿颊香。”如今，“三高”、肥胖等问题普遍存在，六堡茶的保健价值更加凸显。在大健康趋势下，六堡茶将大有可为。

传承：匠心传承古法　创新发展产业

20 世纪 90 年代，六堡茶产业一度陷入低谷，只余少数茶农和企业默默守正创新，传承着这杯千年古茶的味道。21 世纪初，六堡茶产业终于迎来了新的转机。随着国民生活水平的提升和黑茶热潮的兴起，六堡茶在国内市场的消费需求开始显现。

所谓“春江水暖鸭先知”，六堡茶市场的复苏很快在企业端得到了体现。韦洁群于 2004 年在梧州市苍梧县六堡镇创办了黑石山茶厂。她坚持“原产地、原茶种、原工艺、原生态、原仓储”，还原出了最古老的自带槟榔香的纯天然手工茶的味道，备受市场欢迎。

近年来，韦洁群和女儿石濡菲（六堡茶制作技艺自治区级代表

With the improvement of Chinese people's living standards, problems such as "hypertension, hyperlipidemia and hyperglycemia" and obesity are common, and the health care value of Liupao Tea is more prominent. Under the general healthy trend, Liupao Tea will be more promising. With the improvement of national living standards and the rise of the black tea craze, the consumption demand for Liupao Tea in the domestic market has begun to emerge.

性传承人）经营的黑石山茶厂以三产联动的模式迅速发展。上游以"公司 + 农户 + 基地"的模式，推进有机茶园建设，发展有机茶种植；中游坚持采用传统工艺纯手工制作六堡茶，形成了定制化、品牌化、高端化的发展路线；下游以合作联营方式，将六堡茶的种植和制作与茶文旅、茶康养、茶研学、茶文创等有机结合，让六堡茶文化焕发出新活力。

在石濡菲看来，全产业链发展模式是市场推动的自然选择，但无论怎样，种好茶、制好茶、卖好茶是必须坚守的核心经营理念。

市场端的崛起也让六堡茶传承更具底气和活力，韦洁群和石濡菲依然坚持采用传统工艺纯手工制作六堡茶，制作过程不加入任何机械化环节。韦洁群说："越是这样的年代，越要坚持传统工艺，老祖宗的东西总得有人传承。"

正是秉着这样的初心，韦洁群设立技能大师工作室，打破技艺祖传形式，广开六堡茶制茶技艺培训班，将祖传的古法心得和改良技术编制成规范的操作手册传授给村民，实施六堡农家茶标准化培训，培养出一批批传统制茶能手。她还带着自己的手艺走进学校，从教育源头着手，让传统制茶技艺得到更大的传播和发扬。

Today, Wei Jiequn and his daughter Shi Rufei still adhere to the ancient technique of pure handmade tea and do not add any mechanized links. "The more we are in such an era, the more we must adhere to traditional craftsmanship because the things of our ancestors must be passed on." It is with this original intention that she set up the Master Studio and opened many training courses on the tea making skills of Liupao Tea, cultivating a group of traditional tea making experts. She also brought her own craft into school, allowing the traditional tea making techniques to be further spread and developed.

此外，韦洁群和石濡菲收集、整理了六堡茶的历史资料、古时制作六堡茶的工具、各种陈年六堡茶和茶化石，专门建成了一个纯手工作坊与文化旅游一体化的六堡茶文化展示馆，并积极挖掘和延续当地传统的六堡茶风俗文化。

六堡茶人以匠心传承古法技艺，六堡茶企业则以创新促进产业升级。20 世纪 50—60 年代，梧州茶厂首创“双蒸双压”和“冷水渥堆发酵”技术，使得六堡茶的陈化周期大大缩短，带动了中国整个黑茶产业的发展。

20 世纪 70 年代初，云南茶厂就曾派人到梧州茶厂学习发酵技术，由此催生了普洱熟茶。广西六堡茶文化研究会在《六堡茶记》一文中指出，六堡茶的发酵茶比普洱茶早 20 多年。

经过 10 多年市场化发展，梧州六堡茶产业形成了集生产加工、市场交易、技术研发、产品展示与茶旅融合等功能于一体的优势特色产业集群。一片小茶叶，推动了第一、第二、第三产业的融合发展。

2022 年 10 月，习近平总书记在参加党的二十大广西代表团讨论时提出：“茶产业大有前途。下一步，要打出自己的品牌，把茶产业做大做强。”总书记的殷切叮嘱给高速发展的六堡茶产业注入了

Makers of Liupao Tea carry on ancient skills with ingenuity, and enterprises of Liupao Tea promote industrial upgrading with innovation. In the 1950s to 1960s, Wuzhou Tea Factory pioneered the "double steaming and double pressure" and "cold water piling fermentation" technologies, which greatly shortened the aging period of Liupao Tea, realized the development and upgrading of Liupao Tea from traditional agricultural form to modern scientific and technological innovation industry.

六堡茶 ◆一盏茗茶，齿颊留香
Liupao Tea ◆A cup of fine tea, leaving a lingering aroma on the teeth and lips

动力和信心。

守正传承，创新发展。如今，六堡茶制作方面“传统工艺”和“现代工艺”双艺齐飞，市场各有偏好。技艺虽有不同，但殊途同归，都是为了做好这碗“红、浓、陈、醇”的茶汤，传承这道流传千年的茶味。近年来，随着“一带一路”倡议的不断推进和《区域全面经济伙伴关系协定》（RCEP）的落地实施，六堡茶作为著名的“侨销茶”，乘势而上，顺势而为，将迎来新的发展征程。

Nowadays, traditional and modern tea making techniques differentiate Liupao Tea into "farmhouse tea" and "factory tea", with different market preferences. Although there are different skills, the purposes are the same, both in order to produce the bowl of "red, strong, aged, and mellow" tea soup and continue the tea flavor that has been passed down for thousands of years. In recent years, with the continuous promotion of the "The Belt and Road" and the implementation of the *Regional Comprehensive Economic Partnership* (RCEP), Liupao Tea will be ushered in a new development journey.

恭城油茶：以茶入馔　药食同源

Gongcheng Oil Tea: tea-infused cuisine, medicine & food in harmony

"非遗"饮食名片

◆ **茶俗（瑶族油茶习俗）**

类别：中国传统制茶技艺及其相关习俗

级别：联合国教科文组织人类"非遗"代表作

时间：2022 年

地域：广西桂林

桂林山水甲天下，恭城油茶格外香。在广西桂林市恭城瑶族自治县，油茶作为一道历史悠久的地方特色饮食，深受当地人的喜爱。

恭城油茶，也叫"炒米茶"，当地人习惯叫"米花

Guilin's scenery is renowned as the best under heaven, and Gongcheng Oil Tea is particularly fragrant. In Gongcheng Yao Autonomous County, Guilin City, Guangxi, Oil Tea, as a specialty with a long history, is deeply loved by the locals. In 2021, Yao Nationality Oil Tea Custom was selected into the fifth list of representative projects of the National Intangible Cultural Heritage. In 2022, it was included in the UNESCO Representative List of Human Intangible Cultural Heritage.

扫码看视频

茶”。不同于其他地方的油茶的做法，恭城油茶是将茶叶、生姜与大蒜等原材料放入铁锅反复捶打后熬煮制成的饮品，可以搭配炒米、麻旦果和炒好的花生等一起食用。一口油茶喝下后，先感受到茶叶的清苦，过后唇齿留香，味浓而涩，涩中带辣，提神醒脑，饱腹感足。2021 年，茶俗（瑶族油茶习俗）入选第五批国家级“非遗”代表性项目名录；2022 年，入选联合国教科文组织人类“非遗”代表作名录。

恭城油茶 ◆恭城油茶喷喷香，既有茶叶又有姜

Gongcheng Oil Tea ◆Gongcheng Oil Tea smells delicious, it contains both tea leaves and ginger

溯源：以茶入馔　源远流长

恭城位于今广西桂林市东南部，于隋大业十四年（618年）建县，因盛产茶叶而名“茶城”，唐武德年间改名“恭城”。清光绪十五年（1889年）版《恭城县志》载：“隋末梁萧铣起兵巴陵，据粤境，始分平乐地置县，曰茶城。唐武德四年平萧铣，置乐州治，改茶城为恭城县。”

关于恭城油茶的起源，民间有多种说法。有一种说法推测恭城油茶起源于唐代。关于唐代民间流行的煮茶法，陆羽著的《茶经》中记载：“或用葱、姜、枣、橘皮、茱萸、薄荷之等，煮之百沸，或扬令滑，或煮去沫……”大意是把茶饼碾成粉末，放姜、葱、橘皮等作料，在茶锅中煮开。唐代把这种饮茶方式称为“吃茶”。恭城地处偏远，因而保留了以茶入馔的遗风，并一直延续至今。

有一种说法则是恭城油茶与从恭城走出去的北宋监察御史周渭有关。据宋史记载，周渭一生廉洁奉公，执法严明，颇有政绩，深受百姓爱戴，恭城县城东至今仍保留着“周渭祠”。《桂林风俗》中记载着一个流传甚广的故事：周渭回家探亲时，将茶叶和生姜一起捶烂，放油、盐后加水烧开煮成茶汤，用于救治因吸入瘴气而病倒的人，后这一吃法流传于民间。

Regarding the origin of Gongcheng Oil Tea, there is a speculation that it originated in the Tang Dynasty, retaining the legacy of tea as food. Another view is that it is related to Zhou Wei, the imperial censor of the Northern Song Dynasty, whose hometown is Gongcheng. When Zhou Wei went home to visit relatives, he pounded tea leaves and ginger together, added oil and salt, and boiled them into oil tea soup, which was used to treat people who fell ill by inhaling miasma. Later, this method spread among the people.

相比这个故事，如今更多人认为，恭城油茶的诞生与迁徙到恭城来的瑶族人密切相关。

据考证，恭城的瑶族人以平地瑶的周姓一族入居最早，始于唐初。此后，瑶族人从湖南、广东、江西等地陆续迁入，形成“南岭无山不有瑶”的局面。瑶族人多住在崇山之中，居住环境潮湿，多有瘴气。山野之地，缺医少药，为了对抗恶劣的自然环境，聪明的瑶族人就地取材，用茶叶、生姜、蒜头等煮水而食，这种茶汤逐渐演变成今天恭城人人爱喝的油茶。

不管是哪个起源，恭城油茶的历史都已相当久远。打油茶、喝油茶，早已成为恭城人日常生活的一部分。在 1889 年以及 1937 年的《恭城县志》中，就有恭城人爱喝油茶的记载。与恭城油茶相关的故事传说有很多，其中最为人津津乐道的是关于“爽神汤”由来的故事。

据说当年乾隆皇帝下江南时，沿途百官大献殷勤，山珍海味无尽献上，吃得乾隆见食生厌。这时，一位来自桂林恭城的御厨想起家乡油茶之功效，就赶制工具，做出恭城油茶奉上御前。乾隆喝后顿时口舌生津，胃口大开，龙心大悦，金口御赐恭城油茶为“爽神汤”。

More people now believe that the birth of Gongcheng Oil Tea is closely related to the Yao people who migrated to Gongcheng. Most of the Yao people live in the mountains, where the living environment is humid and the miasma is heavy. In order to fight against the harsh living condition, the wise Yao people used local ingredients and boiled water with tea, ginger and garlic, which is very useful for dispelling dampness and eliminating miasma. It was passed down from generation to generation and gradually evolved into the Oil Tea that everyone in Gongcheng likes to drink today.

技艺：恭城油茶慢慢敲

恭城油茶是恭城瑶族饮食文化当中的精华，也是当地独具特色的文化标志。不同于其他茶类，油茶不仅仅是一杯饭后消食解腻的饮品，更是餐桌上当之无愧的主角。

瑶族油茶习俗自治区级代表性传承人、恭城瑶族自治县油茶协会秘书长周黎维介绍，恭城油茶的用料及制作方法均十分讲究，恭城油茶的特色在于“打”，故其茶味与其他地方用沏、泡、煮、煎出来的茶味大不相同。食材的用量、敲打的力度和时间的把控，并没有统一的标准，主要依靠个人的经验和悟性。

（1）准备原材料

恭城油茶是以大叶绿茶为主料，一般选用清明、谷雨两个时节采摘的茶叶。采摘时，叶梢部位的枝条也要一并取下，采用半发酵的技艺，经过炒青、揉捻、炒干、渥堆发酵、摊晾与存储等程序制成。在这个过程中，发酵的技艺是关键。

茶叶的多少视喝油茶的人数而定，一般一人一餐用 20—30 克茶叶就足够了，生姜用量为 15—20 克。其他配料有花生、蒜头、葱花、盐等。

Gongcheng Oil Tea's characteristic lies in "beating", far from the flavor of "brewing", "soaking", "boiling" or "frying" in other places. Gongcheng Oil Tea is mainly made of Daye black tea, which is usually picked in Pure Brightness and Grain Rain. When picking, the leaves on the top of the branches should also be taken down, and then partially fermented techniques are used, such as fixing, rolling, baking, pile-fermentation, spreading to dry and storage. The amount of tea depends on the number of people who drink Oil Tea at a time. Other ingredients include peanuts, garlic, raw green onion and salt.

（2）选取专用茶具

制作恭城油茶要用专用茶具，俗称“油茶三件套”：一个用生铁铸成的专用茶锅，口径大多为33厘米，有柄有嘴，比一般的铁锅小些，但要厚上两三倍，这样的铁锅散热比较慢，耐火耐敲；一把“7”字形的茶槌，一般用原木制成，长的一边作为手柄，另一边用来打碎茶叶和原材料；一个竹编的茶叶隔，用来分离茶汤和茶叶残渣。

油茶三件套 ◆茶锅、茶槌和茶叶隔

Oil Tea trio ◆Tea pot, tea mallet, and tea partition

（3）打油茶

所谓“恭城油茶慢慢敲”，“敲”是恭城油茶制作的点睛之笔。将茶锅烧热，放少许油，将茶叶、生姜、花生、蒜头放入锅内稍炒，随即用茶槌将茶叶、生姜等反复捶打捣烂。恭城油茶制作有“一打二擂”的说法，不仅要轻轻捶打，还要用力碾压，捶至锅底黏上一层黄绿色的锅焦，再加入适量的热开水，熬开出味。

恭城当地有句老话：“打油茶，无他法，只要水滚锅头辣。”95℃以上的热水，150℃以上的热锅，锅下保持旺火，做出来的油茶才能汤色黄爽，香气扑鼻。

（4）出锅

碗中加入炒米、麻旦果、葱花、香菜末等配料，再根据个人口味，放入适量食盐调味，用竹编的茶叶隔把茶水滤出装碗，一碗浓酽喷香的油茶即做好了。

制作好的油茶，其色泽以黄绿色为佳，闻之茶香中夹杂着葱花、香菜的香味，入口绵滑，微辣中带有涩味，随后回甘回甜，令人回味无穷。

周黎维介绍，瑶族人打油茶一般连做四锅，因为第一、第二锅的油茶偏于苦涩，色泽也欠佳，到了第三、第四锅的油茶才是最好

First, heat the pot, put a little oil, and then put the cleaned tea leaves, ginger, peanuts and garlic into it and fry them slightly. After that, beat the tea leaves and ginger repeatedly with the tea mallet. Until the bottom of the pot sticks to a layer of yellow-green tea crust, stop beating and add appropriate amount of hot water, boiling out the flavor. Add rice flakes, madanguo, chopped green onions, chopped coriander, and other ingredients to the bowl. Then, season with an appropriate amount of salt according to personal taste. Filter the tea with a bamboo-woven tea sieve and put the soup into a bowl. A bowl of rich and fragrant Oil Tea is made.

刚出锅的恭城油茶　◆色香浓郁，入口绵滑
Freshly cooked Gongcheng Oil Tea　◆Rich in aroma and texture, smooth and velvety in the mouth

的，所以才会有“一杯苦，二杯涩，三杯四杯好油茶”的顺口溜。

恭城油茶具有消食健胃、提神醒脑、祛湿避瘴、延缓衰老等功效。唐代著名医学家、药物学家陈藏器在《本草拾遗》中写道：“茶为万病之药。”茶叶中含有丰富的锌、锰、镁等微量元素，常用可补充人体所缺的微量元素，茶多酚具有较好的抗氧化作用，茶碱可用来调理全身；花生富含脂肪、淀粉及蛋白质等营养成分，能够补充能量、消食健胃；生姜富含姜黄素，能够解表散寒、止渴消毒；配以葱白，可以通阳顺气，祛除寒湿；蒜头有杀菌的功效。“打油茶”的过程中将这些食材捣烂，再用水熬煮，进一步强化了药食同源的作用。

恭城如今被评为“中国长寿之乡”，许多人认为这与恭城人爱喝油茶的习俗不无关系，恭城油茶也因被人们誉为“长寿密码”之一而声名日隆。

传承：“油茶文化”打天下

在千百年的岁月中，油茶已融入恭城人的生命里。在恭城，可以说是男女老幼都离不开油茶。早晨，恭城人在“打油茶”的节奏中开启了一天的生活，有的家庭甚至一日三餐都离不开油茶，油茶

Gongcheng Oil Tea has the effects of promoting digestion, refreshing the mind, dispelling dampness, eliminating miasma and delaying aging. Gongcheng is now rated as "the hometown of longevity in China", which is related to the custom that local people drink Oil Tea all the year round. Gongcheng Oil Tea is also known as the "longevity code".

早餐、油茶泡粥、油茶米粉……正应了恭城人常说的一句话："恭城土俗，油茶泡粥，祛瘴防疫，全身舒服。"

这种对油茶的喜爱深入恭城人的骨髓，浸润到了他们的日常点滴，在婴儿满月宴、婚嫁、贺寿、祭祀等场合，油茶可以说是无处不在，形成了特定的茶礼、茶规与茶俗。

恭城瑶族文化研究协会会长莫纪德介绍，在恭城栗木镇，过去在婚嫁、生日酒事中，中午主家要打一餐油茶待客。在观音乡上五排村，男方相亲时除了要带上礼品外，最不能少的就是"打油茶"的东西，包括茶叶、生姜及一些喝油茶的小吃。在恭城，瑶族人有用油茶待人接客的习俗，只要有客人来，不分早晚，随时煮好油茶奉客。按当地习俗，客人喝油茶一般不能少于三碗，这叫"三碗不见外"。

"莫讲瑶家礼信差，进屋就喊喝油茶。油茶好比仙丹水，龙肉煮汤不如它。"在千百年的传承中，恭城油茶成为人们维系情感的重要纽带，成为人们交朋结友的重要媒介。

桂林油茶美食协会会长余佳骏介绍，最初油茶只是作为早餐或夜宵在路边小摊上售卖，常喝油茶的人也以恭城人为主。20 世纪 80 年代，桂林市各县的市场经济日渐活跃，恭城陆续出现了一些油茶

In Gongcheng, it can be said that every household, men and women, old and young, are inseparable from Oil Tea. In the morning, people start a day's life in the rhythm of "beating Oil Tea", and some families can not even eat three meals a day without Oil Tea. This love for Oil Tea goes deep and infiltrates into the daily life of Gongcheng people, no matter it is a baby's first month feast, wedding, birthday celebration, or a sacrificial rite. Therefore, it forms specific tea etiquette, tea rules and tea customs.

店。随着越来越多的恭城人来到市区工作，恭城油茶店也在桂林市陆续出现。

进入 21 世纪后，桂林市乃至广西区内的油茶店越来越多，油茶以其健康属性、实惠价格得到越来越多人的认可。

随着油茶餐饮行业队伍的发展壮大，恭城制定了《恭城油茶制作技术要求》和《恭城油茶服务质量规范》两个标准，推动油茶餐饮业的健康发展。

在餐饮食品化的趋势下，油茶制造业近年来也发展迅速，成为油茶产业的另一个支撑板块。浓缩油茶、袋装油茶、油茶伴手礼等油茶系列产品相继被研发并推向市场，越来越多的油茶产品在线下门店、电商平台等走俏。

餐饮及食品业态的不断发展壮大，延伸了油茶的产业链条，不仅带动村民种植茶叶、生姜、红薯、芋头，制作麻旦果、炒米、糕点等，还推动电子商务、物流、服务等产业共同发展，促进第一、第二、第三产业融合发展。

油茶产业发展的同时也助推了瑶族油茶习俗这一“非遗”文化的传播。周黎维近年来通过经营恭城油茶文化体验中心、参加展会、进校讲课等方式，不断输出和传播恭城油茶制作技艺和文化，让更

Under the trend of catering productization, the oil tea manufacturing industry has also developed rapidly in recent years, becoming another supporting sector of the oil tea industry. Concentrated Oil Tea, bagged Oil Tea, Oil Tea hand gifts and other Oil Tea series products have been developed and pushed to the market. The continuous development and growth of catering and food industry have extended the industrial chain of Oil Tea, promoted the integration of the primary, secondary, and tertiary industries, and also promoted the dissemination of the Yao ethnic Oil Tea custom, which is an intangible cultural heritage.

多人了解瑶族的“长寿密码”。

当地政府也大力打造“油茶文化”，通过举办油茶文化节、支持彩调剧《一品油茶七品官》等喜闻乐见的方式，挖掘、展示恭城油茶的独特文化和历史底蕴。2019 年，在第九届桂林恭城油茶文化旅游节上，2019 名瑶族同胞在恭城油茶小镇创造了“最多人一起打油茶”的吉尼斯世界纪录。

现如今，茶俗（瑶族油茶习俗）已入选联合国教科文组织人类“非遗”代表作名录，恭城油茶形成了餐饮、食品、文旅、康养相辅相成的融合发展之路，“油茶经济”日渐兴起。油茶已成为恭城的一张饮食名片，承载着瑶族悠久而璀璨的文化，并渐渐走出大山，走向世界。

The local government has made significant efforts to preserve and promote the Oil Tea culture. Nowadays, Gongcheng Oil Tea has formed the industrialization system of catering, food, cultural tourism and health care, resulting in a gradual increase in the oil tea economy. Oil Tea has become a signature delicacy of Gongcheng.

丹泉酒：优粮好曲　天然洞藏

Danquan Liquor: made from fine grains and excellent yeast, naturally aged in caves

"非遗"饮食名片

◆ **丹泉酒酿造技艺**

类别：传统技艺

级别：自治区级"非遗"代表性项目

批次：第五批

地域：广西河池

十里洞藏，丹泉飘香。丹泉酒产自广西河池市南丹县，其以糯红高粱为本，取凤凰山脉天然山泉，加百草秘制大曲，藏于"洞天酒海"之中，经时光酿造，终得一瓶酱香浓郁、幽雅细腻、醇厚丰满、回味悠长的丹泉

Danquan Liquor is produced in Nandan County of Guangxi. Its raw material is glutinous red sorghum, with the natural mountain spring taken from the Phoenix Mountain Range and distiller's yeast in which all kinds of herbs are added. And then the liquor is stored in the "Dongtian Jiuhai" caves. As time goes on, finally Danquan Liquor is obtained.

扫码看视频

"洞天酒海" ◆世界最大的天然藏酒洞

"Dongtian Jiuhai" ◆The world's largest natural wine storage cave

酒。2014 年，丹泉酒酿造技艺入选广西第五批自治区级“非遗”代表性项目名录。

溯源：历代传承 千年不衰

广西河池南丹县位于广西西北部，地处云贵高原南麓，气候宜人，风景秀丽，是块蕴藏着“长寿密码”的宝地。据乾隆十九年（1754 年）版的《庆远府志》记载，南丹因盛产朱砂，又地处南方而得名。早在明代，著名地理学家徐霞客游览南丹之后，就在其游记中留下了“桂西第一奇胜”的记载。

所谓“好山好水酿好酒”，丹泉酒便产于南丹这块宝地。丹泉酒始于秦汉，兴于唐宋，盛于明清，历代传承，千年不衰。相传，汉武帝征服夜郎国之后，夜郎王为向汉武帝上表臣服之意，特命人用当地炼丹泉水和五谷杂粮酿造美酒，并取名“丹泉”，进贡汉武帝。因醇厚绵甜，丹泉酒随即被定为宫廷御宴酒，更被用于祭祀昆仑山西王母。丹泉酒可谓盛极一时。

唐代诗人曹唐曾写《汉武帝于宫中宴西王母》一诗：“鳌岫云低太一坛，武皇斋洁不胜欢。长生碧字期亲署，延寿丹泉许细看。剑佩有声宫树静，星河无影禁花寒。秋风袅袅月朗朗，玉女清歌一夜

Danquan Liquor which began in the Qin and Han Dynasties, flourished in the Tang and Song Dynasties, reached its height in the Ming and Qing Dynasties, was produced in Nandan, passed down through the generations for thousands of years. According to legend, after Emperor Wudi of Han conquered the ancient kingdom of Yelang, King Yelang specially ordered people to use local alchemy spring water and grains to make a kind of fine liquor in order to show submission to Emperor Wudi of Han, named it "Danquan", and paid tribute to Emperor Wudi of Han to enjoy it. Because of its mellowness and sweetness, it was immediately designated as a palace banquet liquor.

阑。”“延寿丹泉”由此而得，名扬四方。

据记载，自宋朝开宝七年（974 年）设立南丹州至清光绪三十一年（1905 年）“改土归流”的近千年，莫氏土司以丹泉酒为租税向百姓收取，年年向朝廷纳贡以示友好。

历经 2000 多年岁月沉淀的酿酒技艺和文化底蕴，让丹泉酒在历史长河中历久弥新。1956 年 5 月，南丹县政府出资收购南丹境内裕丰、朱林记等几家知名老酒坊，成立南丹县地方国营酿酒厂，早期主产米香型白酒。改革开放后，其厂名更改为国营南丹县酒厂，后因经营不善而停产。2003 年，国营南丹县酒厂改制成了民营企业，组建成立广西丹泉酒业有限公司，注入资金 5 亿元进行迁址扩建，购进先进的生产、包装、检验设备，开启了重建丹泉酒业的新历程。

若以云贵高原为中轴线将地图对折，南丹县所处位置刚好与贵州仁怀市重合。二者高度相仿的地貌与气候特征，形成了相似的酱香型白酒酿造生态环境。改制后的广西丹泉酒业有限公司聘请了茅台镇、古蔺县等地专业的酿酒技术人员进行酱香型白酒的研制。在传承古法酿造技艺的基础上，广西丹泉酒业有限公司成功研制出丹泉洞藏系列白酒，形成了以酱香型为代表，涵盖浓香型和米香型的

In May 1956, the Government of Nandan County made the acquisition of several well-known old distilleries, and established a local state-owned distillery in Nandan County. After the reform and opening up, the name of the factory was changed to the State-owned Distillery of Nandan County, but later stopped production due to poor management. In 2003, it was restructured into a private enterprise and Guangxi Danquan Liquor Industry Co., Ltd. was established with an injection of 500 million RMB for relocation and expansion. Advanced production, packaging, and inspection equipment was purchased, marking the beginning of a new process of reconstruction.

丹泉飘香　◆好山好水酿出的好酒

Fragrant Danquan Liquor　◆Fine liquor brewed from excellent mountain and spring water

产品生产格局，如今已发展成为广西最大酒企。

技艺：古法酿酒　天然洞藏

得天独厚的自然环境与历久弥新的酿造工艺，成就了丹泉酒的高品质。这其中少不了技术人员的坚守和付出，丹泉酒酿造技艺自治区级代表性传承人、丹泉酒业技术总工程师覃建良便是其中一员。

覃建良介绍，“活性洞藏”“以当地本草入曲”是丹泉酒区别于其他白酒酿造工艺的显著标志。

在原料方面，丹泉酒严控品质大关。粮为酒之肉。丹泉酒坚持选用当地生态基地培育的优质糯红高粱，这些高粱严格按照生态要求进行种植，不施化肥，不使用农药，完全使用农家肥，确保了酿酒对原粮品质的需求。

水为酒之血。丹泉酒的酿酒用水取自凤凰山脉下的活性山泉水，其富含硒、钙、铁等多种有益于人体健康的矿物质及微量元素，品质达到饮用泉水的标准。

在具体工艺方面，丹泉酒严格遵循“12987”（即 1 年周期、2 次投料、9 次蒸煮、8 次循环发酵、7 次蒸馏取酒）及“四高三长”（即高温制曲、高温堆积、高温发酵、高温蒸馏，制曲时间长、生产周

The "unique cave" and "secret distiller's yeast" distinguish Danquan Liquor from other liquor brewing processes. In terms of raw materials, grain is the "flesh" of liquor. Danquan Liquor Company insists on selecting high-quality red glutinous sorghum cultivated by local ecological bases. Water is the "lifeblood" of liquor. The brewing water for Danquan Liquor is sourced from the active mountain spring water beneath the Phoenix Mountain Range. The water is abundant in selenium, calcium, iron, and other minerals and trace elements beneficial to human health, and its quality fulfills the criteria for drinking spring water.

期长、储存时间长）的酿酒工艺，以糯红高粱、小麦为原料，经制曲、投料、蒸煮、循环发酵、蒸馏取酒、陈酿洞藏与精心勾兑等多个工序酿造而成。

（1）制曲

曲为酒之骨。在丹泉，世代传承的匠人每年都遵循酿造传统——端午踩曲，重阳投料。每年端午节开始，雨季来临，空气湿度大，温度高，微生物繁殖旺盛，在这种气候下，由小麦制得的高温酒曲自然发酵，温度高达 63℃以上，高温酒曲断面呈金黄色，酱香味浓郁。

同时，丹泉匠人在制作酒曲的过程中因地制宜地融入了 30 多味广西当地中草药，用这样的酒曲酿酒可弱化酒的焦煳味和苦味，使酒的酱香味更为纯粹，从而使丹泉酒具有区别于黔川酱香型白酒的独特风格。其秘制酒曲已获得多项国家专利。

（2）投料

丹泉酒的酿造分 2 次进行投料：第一次在重阳节投料，称为“下沙”；第二次投料在重阳后的 1 个月，称为“糙沙”。

（3）蒸煮

在下沙、糙沙环节各蒸煮 1 次，第三次蒸煮得到熟糟，得到熟

In terms of specific processes, Jiang-aroma Danquan Liquor strictly follows "12987" (1-year period, 2 adding ingredients, 9 steaming, 8 cycle fermentation, 7 distillation) and "four highs and three longs" ("four highs" are high-temperature in making distiller's yeast, high-temperature in stacking, high-temperature in fermentation, and high-temperature in distillation; "three longs" are long making time of distiller's yeast, long production cycle and long storage time) brewing process, using red glutinous sorghum and wheat as raw materials and going through distillation's yeast making, adding ingredients, steaming, circular fermentation, distillation, store in caves, careful blending and other processes.

糟后经摊凉、撒曲、堆积、下窖、封窖发酵、开窖取醅和上甑蒸酒等工序共计 6 个轮次的循环，每个轮次中都有 1 次蒸煮，整个过程共有 9 次蒸煮。

（4）循环发酵

酱香型白酒的发酵在蒸煮下沙时发酵 1 次，混合糙沙蒸煮时发酵 1 次，得到熟糟后经摊凉至上甑蒸酒等工序的 6 个轮次的循环，每个轮次中都有 1 次封窖发酵（每次加曲入窖发酵 1 个月），整个过程共有 8 次循环发酵。

酿造第三轮开始不再投入新的粮食，醅内淀粉含量较高，随着发酵轮次的增加，淀粉会被逐步消耗，直到第八次发酵结束后，醅内的淀粉含量在 10% 左右。

丹泉酒　◆经过多道工序方能酿出的美酒
Danquan Liquor　◆Fine liquor produced by many processes

（5）蒸馏取酒

酱香型白酒第一、第二次蒸煮时不取酒，从第三次蒸煮才开始取酒，共取酒 7 次。35—40℃的流酒温度，更易于香味物质的提取，令酒体具有酱香型白酒的显著风格。

而每次取酒的味道也不一样：第一次取出的酒叫作“生沙酒”，这种酒口感生涩，略带生粮味和酸味；第二次取出的酒叫作“回沙酒”，这种酒酒味醇正，有点涩；第三、第四、第五次取出的酒叫作“大回酒”，这种酒酱香突出，酒味醇厚，酒质丰满，出酒率高；第六次取出的酒叫作“小回酒”，这种酒酱香明显，余味长，有焦香；第七次取出的酒叫作“尾糟酒”，这种酒味道醇和，焦香味重，微苦。

（6）陈酿洞藏

好酒需要“三分酿、七分藏”，刚酿出的新酒较为辛辣，还不能称为佳酿，尚需封存，经岁月的积淀，让时间赋予其醇厚绵长的独特口感。丹泉酒的专属天然藏酒洞“洞天酒海”是在天然溶洞基础上开发而成的，该溶洞形成于 3.5 亿年前，占地面积约为 13 万平方米，被“大世界基尼斯之最”收录为“世界最大天然藏酒洞”。洞内年平均气温在 16—17℃，湿度达 85%—95%，将酒藏于洞内，

In the first and second steaming the liquor is not taken out, and it is taken out from the third steaming with 7 times in total. The flow temperature of 35-40℃ allows the macromolecular Jiang-aroma bouquet substances to be extracted more fully, making the liquor more distinctive in the style of Jiang-aroma liquor. The new liquor that has just been brewed is spicy and sharp and can not be called good liquor. It still needs to be sealed and stored for a long time. After more than 3 years of active storage of "constant temperature, wind, humidity and oxygen", it is possible to turn "rigidity" into "softness".

可加快酒体产酯生香。

山里藏好酒，洞中日月长。丹泉酒历经 3 年以上“恒温、恒风、恒湿、恒氧”的活性洞藏，方能化“刚”为“柔”，形成绵柔甘润、回味悠长的独特口感。

（7）精心勾兑

丹泉酒经 8 次循环发酵、7 次蒸馏取酒后，分不同轮次放入陶缸中，贮存 3 年以上再进行勾兑。勾兑时，需要对不同轮次、不同香型、不同年份的酒进行取样、品尝，按比例调制在一起，反复品尝、调制，直至与标准样一致，这就需要勾兑师有丰富的品评经验和责任心。

一瓶丹泉酒经过 1 年的生产周期，再经 3 年以上的活性洞藏，加上勾兑存放的时间，平均酒龄至少 5 年才能出厂销售，可谓是凝结时间的艺术，集天地日月精华于一身。

传承：酒道匠心　创新发展

“优粮、活水、好曲”是丹泉酒不凡品质的基础；千年传承的古法酿酒工艺和独一无二的“活性洞藏”，则是丹泉酒独特口感的保障。正是在原料、工艺、贮藏等每一个环节选用最严苛的标准，历

After eight fermentations and seven distillations, Danquan Liquor is put into a clay pot in different rounds, stored for more than 3 years and then blended. When blending, it is necessary to sample and taste the liquor of different rounds, different bouquet types and different years, then mix them together in proportion, and repeatedly taste and modulate until they are consistent with the standard sample. After a one-year production, a bottle of Danquan Liquor is stored for more than three years. Plus the storage time in the process of blending, the average age of the liquor commercially available is at least five years.

经岁月积淀，才成就了今天的丹泉美酒。

此外，以市场为主导的产品研发是丹泉酒能快速崛起的根基。在覃建良的带领下，丹泉酒业先后开展了食品加工、生物工程、发酵工程、白酒工艺、白酒勾调、白酒检验、白酒分析等技术研究。覃建良主持的科技课题《高档酱香型丹泉酒的研制》，通过了广西壮族自治区科学技术厅的成果鉴定，该成果在广西属首创，成果达到同类技术国内先进水平。同时，通过与国内大学与研究机构长期合作，丹泉酒业在打造洞藏优势、形成独特口感等方面，精益求精，持续攻关，力求打造核心品质优势。

作为“非遗”技艺传承人，国家一级品酒师、酿酒师，覃建良注重技艺传承、匠心精神传承。他主持举办白酒酿造技术培训班，培训内容涉及白酒节能降耗酿造技术、白酒勾兑技术、白酒包装技术、白酒品评方法与技巧等。

“一群人，一辈子，只做一件事”，这是深入丹泉人心里的一句话。正是这种专注、极致的做事风格，让这家曾经濒临倒闭的国营县级小酒厂在改制后的 10 多年里实现蜕变，丹泉酒以其独特品质被誉为“广西茅台”，成为广西酱酒第一品牌。

2019 年，广西提出“振兴桂酒”战略，河池市也在“十四五”

"Excellent grain, flowing water, and good distiller's yeast" is the basis of the extraordinary quality of Danquan Liquor, and the ancient liquor-making technology handed down for thousands of years is the guarantee of the unique taste of it. It is because of the most stringent standards in the links of raw materials, processes and storage that Danquan Liquor is achieved. In addition, market-oriented product research is the foundation for the rapid rise of Danquan Liquor.

期间把酒业列入百年产业行列。乘此东风，广西丹泉酒业有限公司在夯实广西本地市场的基础上，也加快了全国化布局的步伐，吹响了广西白酒品牌“走出去”的号角，成为“振兴桂酒”的领头羊。

同时，广西丹泉酒业有限公司积极促进产业融合，推动乡村振兴。其采取“公司 + 景区 + 合作社 + 农户”模式，打造了丹泉特色生态酿酒庄园；依托丹泉酒业厂区、喀斯特地貌景观和酿酒文化建成“洞天酒海”景区；建设集高端白酒酿制、休闲度假旅游、世界酒文化展示与电子商务于一体的丹泉小镇。

一滴丹泉涌，酒香飘四方。不到 20 年时间，广西丹泉酒业有限公司就发展成为广西规模最大的优质白酒生产基地。这为“非遗”饮食产业化崛起提供了一条可供借鉴的新路。

山中藏好酒 ◆“活性洞藏”为丹泉酒的独特口感提供了重要保障

Treasured liquor, secluded in the mountains ◆The "unique active cave storage" provides an essential guarantee for the distinctive taste of Danquan Liquor

2020年，国家食品质量监督检验中心经过长达1年的科学实验分析，发布《丹泉白酒风味特色分析报告》。该报告指出，丹泉洞藏酱香型白酒口感比茅台镇酱香型白酒更柔和，比古蔺县酱香型白酒更醇厚，并形成了独特的风格，南丹县是继茅台、古蔺之外的又一个高端酱香型白酒产区。同年，经《中国国家地理》杂志认证，南丹县被誉为中国高端酱香型白酒南部产区。

“酒海洞藏，丹泉飘香，一山酒气，天下无双。”传承千年的丹泉酒，正以“黑马”之势异军突起，以独特品质惊艳市场，重构中国酱香型白酒产区新格局。

In less than 20 years, Guangxi Danquan Liquor Co., Ltd. has developed into the largest high-quality liquor production base in Guangxi, and paved a new path for the rise of intangible cultural heritage catering industrialization.

龟苓膏：引药入膳　养生圣品

Tortoise Jelly: health-preserving delicacy with medicinal effect

"非遗"饮食名片

◆ **龟苓膏配制技艺**

类别：传统技艺
级别：国家级"非遗"代表性项目
批次：第五批
地域：广西梧州

"金龟神灵益寿长，银花解毒消疮疡。黄芪防风土茯苓，清热祛湿养颜方。熬膏合入凉粉草，清爽可口美名扬。"龟苓膏是广西梧州市的一种特产，传统龟苓膏的制作以龟和土茯苓为主料，现代经过配方优化，将土

扫码看视频

Tortoise Jelly is a specialty of Wuzhou City, Guangxi Province, and also a traditional medicinal diet with a long history. In 2008, Wuzhou Tortoise Jelly was recognized as a national geographical indication protection product. And in 2021, the production techniques of Wuzhou Tortoise Jelly was selected into the fifth list of representative projects the National Intangible Cultural Heritage.

茯苓改为茯苓，再搭配多种中药材熬炼而成。成品呈黑色，微带透明，入口嫩滑而清爽，味道微苦而甘甜，具有清热祛湿、滋阴补肾、养颜提神等功效，备受人们喜爱。2008 年，梧州龟苓膏获得地理标志产品保护；2021 年，龟苓膏配制技艺入选第五批国家级“非遗”代表性项目名录。

龟苓膏　◆兼具口感与养生的食疗圣品

Tortoise Jelly　◆A medicinal food with both delicious taste and health benefits

龟苓膏 ◆膏体软弹，清爽可口

Tortoise Jelly ◆Soft and springy paste, refreshingly delicious

溯源：药食同源　清热解毒

梧州古称“广信”“苍梧”，是一座有着 2200 多年建城历史的古老城市，是岭南文化的重要发源地之一。其地处亚热带，气温高，雨量多，人们容易患上湿热之症，因此当地人熬制中草药清热解毒的传统由来已久，而龟苓膏正是其中的一个典型代表。

关于龟苓膏的起源，民间有很多说法，在《梧州龟苓膏》一书中收录的相关民间传说就有近 20 个，时间横跨秦代到民国时期，显示了龟苓膏在当地深远的影响力和悠久的历史渊源。其中，流传较广的有诸葛亮南征求妙方、神奇龟苓膏救明军等故事。

据说三国时，蜀汉皇帝刘备新丧，南方诸郡趁乱起兵叛变。蜀国丞相诸葛亮亲自出马平乱。诸葛亮南征时驻军于苍梧郡，当时蜀军将士多为北方人，初到南方水土不服，大多数将士上吐下泻，严重影响战斗力。诸葛亮很着急，急忙找当地人问个究竟。当地人说，当地气候湿热、多瘴气，于是便献上妙方，以当地特产龟、土茯苓熬汤饮用。诸葛亮令军士一试，果然功效如神，大部分将士痊愈了。后来，诸葛亮为这个妙方取名为“龟苓膏”。

此外，也有一些资料显示，龟苓膏起源于一次偶然入药。中国文联出版社出版的《骑楼城的故事》记述了这样一个传说：明朝时

Wuzhou is located in the subtropical zone, with high temperature and heavy rainfall. Because it is easy to suffer from dampness and heat, the local people have a long tradition of boiling Chinese medicinal herbs served as antipyretic and antidote. Many stories talk about the origin of Tortoise Jelly. There are nearly 20 related folk stories in *Wuzhou Tortoise Jelly*, which contains from the Qin Dynasty to the Republic of China, showing the far-reaching historical influence in the local area. Among them, there are many popular stories such as Zhuge Liang asked for a prescription as pacifying south, and magic Tortoise Jelly saved the Ming Army.

一个名叫李三的人因染上热毒、湿毒，自行治疗时偶然发明了龟苓膏，从此，龟苓膏的配方开始固定下来。

到了明末清初，龟苓膏开始在岭南一带流行。梧州市历史文化研究会副会长彭志创介绍，明朝末期，明军残部南逃至梧州，由于北方官兵对梧州降雨量大、气温高的气候极不适应，众官兵纷纷染上热毒、湿毒，引发各种疾患，军中医官束手无策。后来，食用了龟苓膏，众官兵的痼疾不治而愈。这是龟苓膏最早的大规模使用和传播阶段。

此后，龟苓膏一举成名。据传，其还成为清宫中的养生食品之一，颇得皇亲贵族们的喜爱。据记载，大约在明末清初时候，梧州民间就有食用龟苓膏的习俗和出售龟苓膏的店铺，可见当时龟苓膏在民间颇受追捧。

龟苓膏最初是作为药品进行开发的，主要在药店售卖。后来，由于市场需求量大，慢慢地就有一些作坊式的生产和加工。人们开始采用沿街叫卖，或者在凉茶铺及固定摊点售卖的方式销售龟苓膏，其也从药品过渡到了日常食品。到了清代及民国时期，龟苓膏由于药性偏凉而不猛，味道微苦而甘甜，入口嫩滑而清爽，已发展成为一种民间特色小吃。这一时期，人们采用前店后坊的经营方式销售

Tortoise Jelly was originally developed as a medicine and was mainly sold in pharmacies. Later, due to the excessive market demand, some workshops gradually processed Tortoise Jelly and sold it along the street or infixed stalls like tea stall. Tortoise Jelly has also turned from medicinal use to daily food. By the Qing Dynasty and the Republic of China, Tortoise Jelly had developed into a special folk snack because of its mild medicinal effect and its tender, refreshing, slightly bitter and sweet taste.

龟苓膏，门店遍布梧州。《梧州市志・文化卷》中是这样记述的："到（20 世纪）40 年代，梧州已开始经营龟苓膏并驰名中外。"

一直以来，龟苓膏都以传统的家庭作坊式生产为主，真正工业化生产则始于 20 世纪 70 年代末。在梧州人的开拓创新下，龟苓膏被研制成便于冲调的粉剂、便于携带的易拉罐装及盒装产品，开启了专业化工厂批量生产，龟苓膏从此走出梧州，畅销海内外。

经过不断地改良和发展，如今龟苓膏的产品形态更为丰富，配方含量更加科学。龟苓膏的传承和发展，见证了梧州人生活方式的变更，承载着梧州的历史文化。

技艺：配方讲究　工艺独特

现代龟苓膏的制作以龟和茯苓为主料，按照一定比例，配以凉粉草、甘草、金银花、蒲公英、罗汉果、槐花、栀子和白芷等多味中药材，经多道传统技艺精制而成。

龟苓膏配制技艺自治区级代表性传承人莫桂坚从事龟苓膏制作行业已有 30 余年，他介绍，龟苓膏配制技艺选料精准、配方讲究、投料有序、工艺独特，制作者对药材质量优劣的识别、材料配比的把控及火候的掌握，都直接影响着龟苓膏的品质。

For a long time, Wuzhou Tortoise Jelly has been mainly produced by traditional family workshops, and the industrial production began in the late 1970s. From then on, Wuzhou Tortoise Jelly went out of local area and sold well both domestically and internationlly. The Tortoise Jelly is elaborately prepared by taking tortoise and Poria cocos as main raw materials and adding various traditional Chinese medicines such as mesona chinensis, liquorice, honeysuckle, dandelion, Momordica grosvenori, Sophora japonica, gardenia, angelica dahurica and so on according to a certain proportion.

（1）选材及前期准备

现在，龟苓膏制作一般使用国家允许的非保护性龟种或已获相关部门批准并取得人工繁育许可的龟种，如黄喉拟水龟等，制作高档龟苓膏时也使用人工养殖的金钱龟。

制作龟苓膏使用梧州当地出产的茯苓，炮制好的金银花的干燥花蕾，植物胶含量高的凉粉草，以及冲洗干净或处理好的蒲公英、罗汉果等其他药材。另外，还需要准备蜂蜜、淀粉、白砂糖等。

制作龟苓膏所使用的部分中药材 ◆引药入膳，提升疗效的关键
Some of traditional Chinese medicinal herbs in Tortoise Jelly ◆The key to enhancing therapeutic effects lies in integrating medicinal herbs into dietary foods

（2）熬制龟苓膏药液

制作龟苓膏最关键的工序是“熬龟”，一般使用的是整龟，其中，龟板能滋阴潜阳、消除虚火，龟肉具有补血和血、滋阴补肾的功效。

将龟洗干净后宰杀，去除内脏，砍成小块，慢火熬烂，将龟胶和蛋白质熬煮出来。然后，再放入提前准备好的茯苓、金银花、蒲公英和甘草等药材，加水一起慢煮，用滤袋过滤渣滓后，得到龟苓膏药材提取液。

（3）熬煮凉粉草

将清洗好的凉粉草放入锅内，加入适量的水熬煮，加入溶解好的草木灰碱，熬煮出胶，再用滤袋过滤取得凉粉草提取液。

（4）制作米浆

大米淘洗后加水浸泡 6 小时，用石磨将米研磨成细腻的浆液，再用过滤袋过滤，确保米浆细腻无粗粒。

（5）勾兑

将龟苓膏药材提取液、凉粉草提取液和米浆按照一定的比例勾兑。其中，凉粉草提取液要趁热加入，边加边搅拌，观察膏液的颜色变化、黏稠度变化，直到搅动阻力加大，热浆液变得适度黏稠、

The crucial step in making Tortoise Jelly is the preparation of the tortoise broth. Firstly, the tortoise is washed and slaughtered, its internal organs are removed, and then cut into small pieces. These pieces are gently boiled to extract the tortoise's gelatin and protein. Subsequently, Poria cocos, honeysuckle, dandelion, liquorice, and other pre-prepared medicinal herbs are added, along with more water, and the mixture is cooked further. Finally, a sieve is used to obtain the medicinal broth. Additionally, for the preparation of mesona chinensis liquid, the cleaned herb is placed in a pot, water is added to cover it, and then a dissolved alkaline solution is added for boiling. Again, a sieve is used to obtain the mesona chinensis liquid.

挂勺均匀为止。

（6）炖煮

将勾兑好的膏液放进锅中慢火炖煮，膏体完全熟化，冷却之后凝固成为龟苓膏。

制好的龟苓膏的色泽呈均匀的黑色，膏体软弹，带有中草药香气，切而食之，入口微苦。吃时可根据自己的口味，加入蜂蜜、椰汁、炼乳或水果等，使之口感更为丰富。

龟苓膏具有清热解毒、拔毒生肌、滋阴补肾、保健养颜、促进新陈代谢和提升人体免疫力等功效，是药食同源的典型代表。据记载，龟主治虚劳失血咯血，咳嗽寒热，补服降火；茯苓有利水渗湿、健脾宁心等功效，在《神农本草经》等我国早期本草著作中均有记载，被中医界誉为除湿之“圣品”，健脾之“要药”，再配以蒲公英、金银花等中药材，使药效加强。《普济方》记载：“千金云：凡欲治病，先以食疗。”龟苓膏正是不可多得的兼具口感与养生的食疗圣品。

传承：坚守古方　创新工艺

龟苓膏是历史悠久的传统药膳，是以梧州人民在长期劳动实践中所形成的独特配方和制作工艺制作而成的一种保健食品。

After washing the rice and soaking them in water for 6 hours, grind the rice into rice paste by a stone mill. After that, filter it with a pocket filter to ensure that the rice paste is fine. Next, the extracting medicinal liquid of Tortoise Jelly, the mesona chinensis liquid and the rice paste are blended according to a certain proportion. The mesona chinensis liquid should be added into the blended liquid while being hot and stirred. Finally, boil the blended liquid. When the liquid is completely cooked and cooled, it solidifies into tortoise jelly.

作为“梧州三宝”之一，龟苓膏早已融入了梧州人的生活。走在梧州街头，至今仍然能看到不少做龟苓膏的老店，屋后是小小的作坊，屋前摆上几张桌椅，便是简单的店铺。围桌而坐，来一碗龟苓膏，既可把它当作清甜的“中药”吃，又可当作略苦的“甜品”进行品尝，在甜与苦的味蕾冲突中，实现了健康与美味的平衡。

在坚守古方传承的同时，梧州人对于龟苓膏的制作工艺，也在不断改良研发，使之与时俱进。莫桂坚介绍，20 世纪 80 年代之前，龟苓膏只是梧州街头的传统小吃，受保鲜技术限制，只能保存一两天，无法远销。20 世纪 80 年代初期，广西梧州市第三制药厂在厂长杨惠堂的组织带领下，传承梧州龟苓膏古方，研制出了可以用开水冲调的龟苓膏粉。1983 年，广西梧州市第三制药厂梧州饮料部成立，并在梧州文化路开了一间售卖龟苓膏的店铺，出售即制即食的传统龟苓膏，取名“双钱”店，吸引了大批游客前来品尝。“双钱”成为梧州对外的一个饮食招牌。

龟苓膏粉则通过专业化工厂批量生产，在广西、广东、香港、澳门等地销售，并出口至东南亚地区，龟苓膏开始名声大振。1993 年，广西梧州双钱实业有限公司研究发明了易拉罐装龟苓膏；2002 年，又成功研制经高温灭菌、不加防腐剂的即食型塑料碗装龟苓膏，

While adhering to the heritage of the ancient prescription, Wuzhou people are constantly improving the production techniques of Tortoise Jelly and keeping pace with the times. In the early 1980s, under the organization and leadership of Yang Huitang, director of the Third Pharmaceutical Factory in Wuzhou City, Guangxi, the ancient formula of Wuzhou Tortoise Jelly was handed down, and a powder that could be brewed into Tortoise Jelly with boiling water was developed, Tortoise Jelly powder realized mass specialized production. These products were sold in Guangdong and Guangxi, Hong Kong, Macao, and even exported to Southeast Asia.

使其产品形态更为丰富，更适合现代人的消费习惯。

重视市场，持续创新，让双钱牌龟苓膏保持着增长活力。“年轻消费者更注重时尚，喜爱‘轻食主义’，现在，我们不仅在口感上对产品进行改良，加入红豆、蜜糖等配料，更在包装上下功夫，推出果冻粒装以及可以吸的龟苓膏系列、丰衣粥食系列、龟苓宝系列、高端龟苓膏系列、‘非遗’传承系列等品种，产品销量持续增长。”时任广西梧州双钱实业有限公司副总经理的莫桂坚介绍。

在龟苓膏工业化生产的发展中，广西梧州双钱实业有限公司不断做大做强，其龟苓膏产品及品牌得到了广泛传播。2011 年，“双钱”品牌荣获“中华老字号”称号。2021 年，广西梧州双钱实业有限公司被认定为国家级“非遗”代表性项目（龟苓膏配制技艺）的保护单位。公司不仅以师传方式传承龟苓膏配制技艺，还致力于传统技艺的深入挖掘、整理、建档、展示、理论研究及数字化保存，为“非遗”代表性项目的传承保护提供不竭动力。

以双钱为代表的龟苓膏企业的崛起路径，展示了食品工业化过程中科技研发及市场开拓的重要性。通过技术创新，实现“非遗”饮食产品标准化、工业化生产，突破时空限制，丰富产品品类，拓展市场空间，是“非遗”饮食企业做大做强的基础。

Since then, Shuangqian Company has successively launched canned Tortoise Jelly and plastic bowl of Tortoise Jelly, which enriched the variety of product and made it more suitable for modern people. Paying attention to the market and innovation, Tortoise Jelly of Shuangqian Company maintains its growth vitality. Now, it not only improves the taste of products by adding red beans, honey and other ingredients, but also works hard on packaging. It also introduces jelly granules and suckable Tortoise Jelly series, porridge series, concentrated Tortoise Jelly series, top-grade Tortoise Jelly series.

近年来，梧州市政府重点打造医药食品等六大产业，营造良好的营商环境和发展环境，深入挖掘“中华（广西）老字号”品牌资源，壮大龟苓膏等传统特色食品产业。智研瞻产业研究院数据显示，2022 年中国龟苓膏市场近 60% 的产品均产自广西梧州。

虽然龟苓膏占据了市场的半壁江山，但长期以来，由于浓重的地域特色和文化差异，龟苓膏的食用习惯更多盛行于广西、广东、香港、澳门以及东南亚各地，其他地区对于这类食品的认知度和接受度还不高，呈现出地域分布不均的特点。从行业角度来看，目前整个龟苓膏行业尚处于导入期。在发展健康休闲食品的大趋势下，龟苓膏产业化仍大有可为。

改良后的龟苓膏　◆吃法多样，口感丰富

The improved Tortoise Jelly　◆Variety of eating methods and rich flavors

传承是根，创新是叶。经过数百年发展与变革，如今的梧州龟苓膏已从药品演变为日常食物，从偏居一隅的地方特产变为远近闻名的养生圣品，被越来越多的人所喜爱。梧州龟苓膏产业的不断发展，推动着龟苓膏文化的对外传播。龟苓膏，这道传统美食展现出中国独具特色的中医药文化和博大精深的饮食文化。

In recent years, Wuzhou government has focused on building six major industries, such as medicine and food, providing a good business development environment. It digs deeply into the brand resources of "China (Guangxi) Time-Honored Brand", and publicizes traditional food such as Tortoise Jelly. According to the data of Zhiyanzhan Industry Research Institute, nearly 60% of the products in China's Tortoise Jelly market in 2022 are produced in Wuzhou, Guangxi.

合浦大月饼：大若圆盘　松软香溢

Hepu Big Mooncake: the soft orange-colored cake is as big as a disk and has the wonderful aroma

"非遗"饮食名片

◆ **合浦大月饼制作技艺**

类别：传统技艺

级别：自治区级"非遗"代表性项目

批次：第八批

地域：广西北海

"小饼如嚼月，中有酥与饴。"合浦大月饼是非常有特色的地方小吃，其制作历史悠久，馅料以五仁、叉烧、糖浆等为主，外形以大饼状为特色，色泽橙黄，口感以皮酥馅饱、松软香甜、油而不腻著称，深受消费者

Hepu Big Mooncake is a very characteristic local food. Its production history is long, shape is characterized by bigness, and taste is delicious. In 2018, Hepu Big Mooncake was listed as a protected product of national geographical indication. In 2020, Hepu Big Mooncake Making Technique was selected into the eighth list of representative projects of Intangible Cultural Heritage at the autonomous region level in Guangxi.

扫码看视频

合浦大月饼　◆风味独特，以大著称

Hepu Big Mooncake　◆Unique flavor, renowned for its size

追捧。2018 年，合浦大月饼获得地理标志产品保护。2020 年，合浦大月饼制作技艺入选广西第八批自治区级“非遗”代表性项目名录。

溯源：千年古郡　薪火相传

广西合浦历史悠久，素有“南珠故郡，海角名区”之称，是古代海上丝绸之路的起点，大河文明及大海文明在这里交汇。早在汉元鼎六年（公元前 111 年）便设合浦郡，唐贞观八年（634 年）改称廉州，一直至明清两代仍为廉州府治。2000 多年的历史沉淀让古郡合浦拥有众多文化遗存，月饼文化便是其一。

我国自古就有祭月和赏月的习俗，合浦大月饼制作技艺的发展历史已有千年之久。据坊间说法，客家人从宋元时期开始迁入合浦，在尚未开发的山区聚居，生活条件十分艰苦。为了在中秋节令能制作月饼来祭祀祖先，合浦客家人就想出了由各家各户凑原料做月饼的办法：有米粮的出米粮，有糖油的出糖油，有鸭蛋的出鸭蛋，有豆子的出豆子，又把凑得的原料集中在一起，推选懂技术的乡亲统一制成大月饼。大月饼做好后，人们在中秋之夜集中祭祀祖先，再按比例把月饼送回各家各户分享过节，几户人围起来一起吃。合浦大月饼就这样诞生了，中秋节吃大月饼的传统也相应地流传了下来。

According to anecdotal accounts, the Hakka people began to migrate to Hepu from the Song dynasty, and lived in undeveloped mountainous areas. In order to make mooncakes to sacrifice their ancestors during the Mid-Autumn Festival, the Hakka families in Hepu came up with a way to make mooncakes. That was to collect raw materials such as rice, grains, sugar, oil, eggs and beans from each family and gathered them together to make large mooncakes. After the big mooncakes were made, they were sacrificed to the ancestors on the night of the Mid-Autumn Festival, and then sent mooncakes splited in proportion back to each family to share. The tradition of Hepu Big Mooncakes has been passed down.

资料显示，合浦大月饼是由宋代“猪仔饼”逐渐发展而成的，清康熙六十年（1721年）修订的《廉州府志》中记载，公元1100年，北宋大文豪苏东坡从海南儋州回京城，途经合浦时与廉州太守张左藏会晤并品尝月饼，事后作诗《留别廉守》相赠：“编萑以苴猪，瑾涂以涂之。小饼如嚼月，中有酥与饴。悬知合浦人，长诵东坡诗。好在真一酒，为我醉宗资。”北宋之时，月饼在民间俗称“小饼”“月团”，诗中的“小饼”指外形像猪仔的月饼。

及至清朝，合浦大月饼逐渐兴起。清道光年间的《廉州府志》中有着“八月望夜赏月亦食大月饼”的记载，如今的合浦月饼小镇展示厅内还收藏着从明清以来制作大月饼的模具，这些模具见证了合浦大月饼生产悠久而辉煌的历史。

到了民国时期，合浦大月饼的制作工艺已经有了很大提高，品种不断增加。当时合浦国强楼、大鼻蔡等饼家已在合浦及周边地区十分有名。在历届岭南中秋月饼品尝大会上，合浦大月饼多次荣获第一名。

历经千年薪火相传，如今的合浦已成为著名的“中国大月饼美食之乡”。合浦大月饼制作技艺日益精进，产品样式繁多，合浦也已成为广西乃至全国闻名的月饼生产基地，合浦大月饼声名远播东

According to the data, Hepu Big Mooncake was gradually developed by the Song Dynasty "piglet cake". In the Qing Dynasty, Hepu Big Mooncakes gradually got popular. By the Republic of China, the production technology of Hepu Big Mooncakes had been greatly improved, and the variety continued to increase. Since the technique of making mooncake was passed down for thousands of years, Hepu has become the famous "Hometown of Chinese Big Mooncake Delicacies". Hepu Big Mooncake making technology is becoming more and more refined with a variety of styles. Hepu has become a veritable mooncake production base in Guangxi and even the whole country, especially in Guangdong and Guangxi.

南亚各国，成为合浦的一张美食名片。

技艺：选材严苛　工艺独特

千年古郡，月饼飘香。合浦大月饼以五仁叉烧月饼为代表，风味独特，以大著称。正是严苛的选材、丰富的配料及独特的工艺造就了合浦大月饼的与众不同。

合浦地理环境优越，盛产橄榄仁、黑毛土猪、白砂糖、蜂蜜等，这些都是制作合浦大月饼的主料。合浦大月饼选材严苛，除了选取当地优质食材外，还选取全国各地的优质食材，只为呈现最优品质。

合浦大月饼制作技艺自治区级代表性传承人、合浦公馆黄记食品有限公司董事长黄庆强介绍，合浦大月饼最大的特色就是料足味美，货真价实。在制作过程中，每一道工序都很细致讲究，要用心投入，坚守品质，才能做好每一个月饼。

（1）前期准备

前期准备工序有叉烧的炒制、糖浆的熬制和五仁的挑选等。

叉烧选材于黑毛土猪的圆柱五花条，加入了合浦特产鸡头荔枝蜜、土酒、五香粉等调料腌制，传统工艺是经文火慢炒逼出肉中水分，几经翻炒后得到的叉烧肉质紧密、香味独特。

Hepu Big Mooncakes which are represented by the fillings of Wuren and Chashao has a unique flavor and is famous for its bigness. It is the strict selection of materials, rich ingredients and unique craftsmanship that make Hepu Big Mooncakes outstanding and distinctive. The production process is unique and exquisite. The preparation of raw materials in the early preparation includes the boiling of syrup, the pickling of Chashao, the selection of Wuren. The later steps include preparing the exterior, mixing the fillings, wrapping the fillings, molding, baking, cooling, packaging and other processes.

制作饼皮的糖浆需提前进行熬制，在煮开的糖浆中按比例加入新鲜的柠檬、菠萝促进蔗糖的转化，后放于瓦缸中存放 2 个月以上，让糖浆变得芳香绵柔。

五仁则包括瓜子仁、核桃仁、芝麻仁、杏仁和橄榄仁。五仁寓意深远，蕴含着五福临门、五子登科等吉祥祝福。黄庆强介绍，五仁的挑选非常讲究，如选用新疆的瓜子仁、云南的核桃仁、广西的橄榄仁等，注重从源头把控出品质量。

（2）后期制作

后期制作工序包括和馅、饼皮制作、包馅、成型、烘烤、冷却、包装等。

制作合浦大月饼所需的食材　◆选材严苛，注重品质

Ingredients for Hepu Big Mooncake　◆Strict selection, emphasizing quality

很多人以为大月饼就是由各种食材堆砌而成的，馅料越足越好，其实不然。如果不注重各种食材的比例和国人的口感喜好，做出来的月饼只会又渣又腻。此外，和馅时讲究手法，让各种食材均匀混合，这样做出来的馅料才香味浓郁、口感松软。

月饼饼皮的制作也有讲究。精选优质小麦粉，加入适量的糖浆、花生油等，搅拌均匀后反复揉捏。这样制作出来的饼皮软糯柔润，延展性好，适合放入各种形状的模具里压制成型。每张饼皮都要做成一样大小，一样厚薄。

将月饼馅料称重，以保证馅料份量一致。用饼皮包好月饼的底层和四周部分，确保馅料都包裹严实。由于五仁容易刺破薄薄的饼皮，包馅时力道要轻柔。

在包好馅料的月饼上均匀撒上面粉，放入模具，用掌心适度发力，快速压制成型。然后手握模具，让模具的顶部、底部、左侧、右侧几个不同方位快速与台面发生碰撞，在重力加速度的作用下，饼面和模具发生分离，一个印有模具图案的月饼便成型了。

到了烘烤环节，要给月饼外皮刷上蛋液，这样烤出来的月饼表皮酥黄美观，吃起来更加香甜。待月饼出炉冷却后，再进行真空包装，一个合浦大月饼便可面市了。

Five kernels (Wuren) include melon seeds, walnuts, sesame seeds, almonds and olive kernels. When mixing the fillings, attention is paid to the skill so that the various ingredients are evenly mixed and the fillings are rich in flavor and soft in taste. There is also a special attention to the making of mooncake crust. After selecting high-quality wheat flour and adding syrup, peanut oil and so on, knead the dough to make the crust soft, ductile, and suitable for various shapes of molds.

合浦大月饼的制作 ◆采用传统工艺，工序细致考究

The production of Hepu Big Mooncake ◆Adopt traditional craftsmanship with elaborate and refined procedures

采用这种传统工艺制出的合浦大月饼，皮酥、油足、味香，一口下去，饼皮松软绵柔、不粘牙，果仁香脆可口，叉烧咸甜适度、肉质紧密易嚼，不同层次感在口中绽放，香而不腻，让人回味无穷。

合浦大月饼馅料搭配合理，营养均衡。五仁馅料富含维生素、不饱和脂肪酸、氨基酸等营养成分。其中，杏仁滋养止咳，橄榄

Weigh the mooncake fillings to ensure that the weight is consistent, and then wrap the bottom and surrounding parts of the mooncake with the crust to ensure that the fillings are tightly wrapped. Sprinkle the stuffed mooncake evenly with flour, put it into a mold, and use the palm to exert moderate force and press it quickly, and a mooncake with a mold pattern is formed. When it comes to the baking process, it is necessary to brush the outer crust of the mooncake with egg liquid, so that the surface of the baked mooncake is crispy and color-oranged and the taste is more sweet. After the cake is baked and cooled, it is vacuum-packed and commercial available.

合浦大月饼　◆馅料丰富，层次分明

Hepu Big Mooncake　◆Rich fillings, layered texture

仁去燥解毒，瓜子仁护心降压，芝麻仁润肠助脾，核桃仁顺气补血，合浦大月饼可谓集精华于一饼。

传承：创新工艺　与时俱进

“月”满合浦，“饼”香万家。在如今“小而美”的饮食风潮下，合浦大月饼以大取胜，反其道而“出圈”。每一块合浦大月饼的背后，都寄寓着“大团圆”的中国传统文化，承载着合浦人对传统制作工艺的执着坚守，见证了合浦人在岁月沉浮中的薪火相传。

如今在合浦，依然有很多老字号月饼作坊坚持手工制作月饼。虽然有些作坊后来升级为了工厂，在烘烤、冷却与包装等工序引入自动化设备进行标准化生产，但是在月饼制作的关键工序仍保持手工制作，在保证月饼口感的基础上，将传统工艺与现代科技相融合。

黄庆强介绍，虽然如今黄家月五仁叉烧大月饼实现了工业化量产，但在原料挑选、和馅、包馅等工序依然坚持手工制作。机器无法完全把原料筛选干净，必须要再次进行人工检查、清理，杜绝细小粉碎的杂物，以保证原料的超高纯净度。另外，只有手工包出来的月饼，才能达到饼皮厚薄均匀、断面皮心分明及馅料口感匀称的效果。

Hepu Big Mooncake material proportion is reasonable and their nutrition is balanced. Wuren filling is rich in vitamins, unsaturated fatty acids, amino acids and other nutrients. Almonds have the effect of nourishing and relieving cough, olive kernels can decrease internal heat and detoxify, melon seeds protect the heart and lower blood pressure, sesame kernels lubricate the intestines and improve the spleen, and walnut kernels replenish blood and smooth qi. It can be described that all the genuine materials make the essence in one cake.

月饼的销售期虽然只有一两个月，但月饼从原料筹备到制作完成往往要经过半年多的时间。一般每年清明节过后，就要开始熬制糖浆，过了农历七月十四，便进入了月饼生产季，经常需要上千人参与制作，才能保障产能供应。在黄庆强看来，这些手工制作工序目前是不能被替代的，否则就会失去月饼原有的口感和品质，这背后正是对品牌匠心的坚守。

在市场化发展过程中，黄庆强也注重对合浦大月饼制作工艺进行创新。例如：在叉烧制作过程中，将传统的炒制改为烘制，腌制好的叉烧在 50—60℃的烘焙房里慢慢烘干，这样制成的叉烧更加鲜香，保质期更长；在五仁等原料加工方面，将传统的烘烤工艺改为先蒸煮、后脱水的工艺，由此制成的馅料吃后不易上火；在烘烤环节，由原先的一次性烘烤改为两次，第一次烘烤至月饼表皮硬化后，再刷上蛋液进行第二次烘烤，这样烘烤出来的月饼表面花纹更加清晰，颜色更加金黄；在保持原有配料的基础上，降低甜度、油脂，或使用木糖醇等代糖产品，使月饼更符合当下的健康饮食理念；等等。严格选择材料，保持手工制作的口感，同时创新工艺，才成就了极具口碑的黄家月五仁叉烧大月饼。

近年来，合浦也不断加大对合浦大月饼品牌的保护和投入，充

Although the mooncakes are sold only for one or two months, it takes more than half a year of preparation and production. Generally, after the Qingming Festival every year, syrup will begin to be boiled, and after the 14th day of the seventh lunar month, the mooncake production season has started. Thousands of people are often required to participate in the production to ensure the supply. Strict selection of materials, maintenance of handmade taste and innovation of craftsmanship contribute to the highly acclaimed Huang Jia Yue Wuren Chashao Big Mooncake.

分挖掘合浦大月饼的美食文化底蕴，打造月饼文化节等，传承并发扬月饼文化和制作技艺。同时，通过建设合浦月饼小镇，实现企业集聚、产业集群，发挥资源共享优势，走出了一条“非遗”产业高质量发展之路。

“人生代代无穷已，江月年年望相似”，文化的传播从未间断，技艺的传承从未停滞，曾经的“南珠故郡”，如今以“中国大月饼美食之乡”的名片，再次屹立在世人面前。跨越千年时光，品着相似的月饼，赏着同一轮圆月，过去与现在在此交会，“非遗”传承由此有了不同的温度与体悟，知来路，识归途！

In recent years, Hepu has also continuously increased its protection and investment in the "Hepu Big Mooncake" brand, fully explored its food culture heritage and created mooncake culture festivals to introduce and promote the mooncake culture and production technique. And through the construction of Hepu Big Mooncake town, enterprise agglomeration, industrial cluster, resource sharing are realized.

壮族五色糯米饭：草木染就　香糯可口

The Zhuang Five-color Glutinous Rice: dyed by plants and herbs, tasty and glutious

"非遗"饮食名片

◆ **壮族五色糯米饭制作技艺**

类别：传统技艺

级别：自治区级"非遗"代表性项目

批次：第三批

地域：广西南宁

"三三寒食染五谷，五彩吉祥又如意"，说的就是壮族五色糯米饭。五色糯米饭，又称"五色饭""花糯饭"等，因糯米饭呈五种不同颜色而得名。每年"三月三"、清明节等节日期间，壮族家家户户都会做五色糯米饭。

The Zhuang Five-color Glutinous Rice, commonly known as Five-color Rice, also known as Flower Rice, is named for its five colors. Every year on the third day of the third lunar month, Qingming Festival and other festivals, every household of the Zhuang people will make Five-color Glutinous Rice.

扫码看视频

五色糯米饭 ◆五彩斑斓，香糯可口

Five-color Glutinous Rice ◆Colorful, fragrant and delicious

五色糯米饭取材天然，其多彩的颜色都是从天然可食的植物中提取汁液染成的。五色糯米饭，观之如一朵盛开的五色花，色泽艳丽，食之能“坚筋骨、益肠胃”，强身健体，寄寓了壮族人民对吉祥如意、五谷丰登的美好追求。2010 年，壮族五色糯米饭制作技艺入选广西第三批自治区级“非遗”代表性项目名录。

Five-color Glutinous Rice uses natural materials, and its colorful glutinous rice is mainly dyed with some natutal and edible plant juices. It looks like a blooming five-color flower with bright color, which can fulfil the bones, benefit the appetite and strengthen the body. It shows the Zhuang people's pursuit of good luck and rich harvest. In 2010, The Zhuang Five-color Glutinous Rice Making Technique was selected into the third list of representative projects of Intangible Cultural Heritage at the autonomous region level in Guangxi.

溯源：纪念特掘和乜掘

我国自古就有用植物汁液与米相浸染后蒸煮食用的传统，早在南朝陶弘景所著的道家著作《登真隐诀》中就介绍了青精饭（乌饭）的做法："以南烛草木煮汁渍米为之。"青精饭，是当时道家修炼养生的重要食物。

到了唐代，很多文人墨客都把青精饭看作食疗养生的佳品，杜甫在诗中就曾提到："岂无青精饭，使我颜色好。"宋代以后，道家的青精饭逐渐为佛家吸收，农历四月初八浴佛节这天，民间以青精饭供佛。如今，我国南方一些地区依然保留着"四月八，吃乌饭"的传统。

在广西，每逢"三月三"、清明节、牛王节（农历四月初八）、端午节等民间传统节日，壮族家家户户都要做五色糯米饭，以祈祷五谷丰登、五福临门、吉祥如意。

壮族五色糯米饭制作技艺历史悠久，壮族民间一直口口相传着"'三月三'掘尾龙拜山"的神话故事。相传古时候，大明山下住着一个老妪，她没有子嗣，后来救了一条受伤的小蛇。小蛇伤好后不肯离去，老妪便对小蛇说："人畜有别，我们人是没有尾巴的，你若想跟我在一起，就要舍弃你的尾巴。"后来，老妪就把小

Since ancient times, there has been a practice of steaming rice with natural plant juice in China. In the Tang Dynasty, many writers and poets regarded black rice as a good food for food therapy. After the Song Dynasty, Taoist Qingjing rice was gradually introduced to Buddhism. In Guangxi, on the third day of the third lunar month, Qingming Festival, Buffalo Festival (April 8 of the lunar calendar), Dragon Boat Festival and other traditional folk festivals, every household of the Zhuang will make five-color glutinous rice to pray for a rich harvest, blessings and good luck.

蛇的尾巴斩掉一小截，于是人们就把小蛇叫作“特掘”（在壮语里，“特”是男的意思，“掘”是短尾巴的意思），把老妪叫作“乜掘”（“乜”在壮语里是母亲的意思）。从此，乜掘就把特掘当成儿子一样养着。

后来，乜掘由于年老去世了，邻居们把乜掘入殓后，忽然风雨大作，一阵狂风把乜掘的灵柩和特掘卷到了大明山的最高峰上，那时正值农历三月初三。后来每逢农历三月初三前后，大明山地区出现大风大雨，当地人就认为是特掘回来扫墓祭拜他的母亲了。

人们为了歌颂乜掘的伟大品质和传颂孝道，便给其立庙，在壮语里将之称为“妲婆庙”或“姥婆庙”，汉语里统称为“龙母庙”，于是乜掘就被尊称为“龙母”。农历三月初三就成为壮族人传统上的扫墓时间。据说，因为特掘以前爱吃他母亲做的糯米饭，当地人就按他身上的花纹颜色做成了五色糯米饭，以此来祭拜龙母和祖先，祈望风调雨顺、五谷丰登。

在悠久的历史变迁及世代传承下，五色糯米饭已成为壮族人在“三月三”等节日的必备美食。明代记录广西风土人情的文人笔记《西事珥》就有“诗：岂无青精饭，使我颜色好。今夷人以社日相馈遗，然又有染作青、黄、赤色以相杂者，谓之五色饭”的记载。据

In the folklore of the Zhuang, there has been a myth. An old woman lived at the foot of Daming Mountain in ancient times. She saved a wounded little snake and raised the snake as a son. After the neighbors buried the old woman, the wind and rain suddenly blew, and a gust of wind swept the old woman's coffin and the snake to the top of the highest peak of Daming Mountain. It was the third day of the third lunar month. The third day of the third lunar month has also become a traditional tomb-sweeping day for the Zhuang people. Because the snake used to like eating glutinous rice, the local people made five-color glutinous rice according to the patterns of the snake.

清代的《上思州志》记载，上巳日（即“三月三”），乡民炊糯为饭，以野树叶渍汁染之作红黄色或乌色，以祀先祖。可见，古时吃五色糯米饭的习俗与如今的习俗相差无几。

此外，在骆越文化中，糯米被当作有灵性的食物，用其制成的五色糯米饭还融合了壮族先民对“五方”“五色”“五行”等图腾观念的崇拜和信仰。壮族人认为，作为祭祖祭神仪式活动中必备祭品的五色糯米饭，能驱山精邪祟。清光绪年间的《归顺直隶州志》即有“清明前后拜扫坟墓，俱染糯米成五色，蒸熟供献”的记载。

如今，壮族人在“三月三”做五色糯米饭的时候，仍然会插上一枝精心挑选的枫树叶，甚至做五色糯米饭剩下的染料渣，人们也要将其撒在房屋外墙下，以求驱邪保平安。现如今，五色糯米饭不仅是壮族传统的祭祀供品，还融入了壮族人的日常生活，成为社交送礼、待客之用的传统美食。

技艺：大自然馈赠的香糯与艳丽

五色糯米饭一般呈红、紫、黄、黑、白五种颜色，不同地区的壮族制作的五色糯米饭的颜色并不完全一致，但制作流程基本相同。

Glutinous rice is used as a psychic food in the ancient Luoyue culture. Five-color Glutinous Rice combines the worship and belief of the ancient Zhuang people on the totem concept of "Five Directions", "Five Colors" and "Five Elements". People believe that the five-color glutinous rice, as a necessary sacrifice to ancestors and gods, can expel evil spirits.

壮族五色糯米饭制作技艺自治区级代表性传承人、南宁市武鸣区文化馆五彩壮乡分馆馆长潘红华认为，要制作出成色漂亮的五色糯米饭，需要从选材、染色和蒸制三方面把好关。

（1）选材

为了保证五色糯米饭的地道口感，潘红华选用的是武鸣当地农户种植的糯米。给糯米染色的汁液均取自纯天然植物，枫树叶可染黑色，一般选用清明前后新长出的嫩叶和枝条提取染色汁液，早在清代编纂的《武缘县图经》就有关于农历三月初三取枫树叶泡汁染饭为黑色的记载；红色汁液取自红蓝草；紫色汁液取自紫番藤；黄色汁液取自栀子、黄花、黄姜等；白色便是糯米本来的颜色。

潘红华介绍，红蓝草和紫番藤其实是同一品种而叶状不同的植物：叶片稍长、背面光滑的是紫番藤，用它煮出来的水是紫色的；叶片较圆、背面有绒毛的是红蓝草，用它煮出来的水是红色的。

（2）染色

浸泡糯米，使其上色是制作五色糯米饭最关键的一步，植物需要泡多久，汁液与糯米的比例如何，糯米要泡多长时间，等等，这些细节都不容出错，否则无法染出成色漂亮的彩色糯米。

提取染色汁液时需注意，除了枫树叶要切碎捣烂后直接泡清水，

The dyes of the Five-color Glutinous Rice of the Zhuang are all pure natural plants. The black dye is taken from maple leaves. The red dye comes from red dye grass, the purple dye comes from purple dye vine, the yellow dye comes from gardenia fruit, yellow flowers or turmeric, and the white is the original color of glutinous rice.

24 小时左右再取汁液外，其他原料如黄花可整束，黄姜、栀子需捣烂，红蓝草、紫番藤需砍成段，用水煮之后直接滤渣取汁。

用染色汁液泡米前，汁液温度应在 38—45℃之间，汁液与糯米的比例为 1.5∶1 为宜，然后将米放进不同的汁液中，浸泡大概 4—8 小时，让其上色。

制作五色糯米饭的染料 ◆取材天然，草木染就

Dyes for Five-color Glutinous Rice ◆Naturally sourced, dyed with plants and herbs

（3）蒸制

蒸制也是一项技术活，染好色的糯米不能混杂，用筛子分别沥干水分后，将不同颜色的糯米放入已装好竹篾网的木甑中。然后，将木甑放入锅中，锅内放少量水，水量以刚齐平木甑底层（木甑脚往上 3—4 厘米）为宜，蒸煮 2—3 个小时即成，要注意头 1 小时用文火，之后需用猛火。

经过高温蒸制的五色糯米饭色彩艳丽，颗粒饱满，米饭裹着植物的清香，如一件美食艺术品，视觉上赏心悦目，味觉上唇齿留香。除了直接品尝外，可将花生和芝麻等炒熟、炒香、舂碎，撒入糯米饭中拌匀；也可根据个人爱好，放入少许葱花、油、盐、腊味等一起炒，炒后的五色糯米饭浓香可口。

五色糯米饭不仅味道鲜美，还营养丰富。用来提取染色汁液的植物具有一定的药用价值：红蓝草有生血作用，清代张志聪所著的《侣山堂类辩》中有“红花色赤多汁，生血行血之品也”的记载；李时珍在《本草纲目》里说，枫叶能止泄益睡，强筋益气力，而用枫叶煮成的青精饭，人食之能坚筋骨、益肠胃等；紫番藤有改善血液循环、抗氧化等作用；黄姜利胆、抗菌、降脂，黄花或栀子有清热凉血等作用。用这些植物提取的汁液浸泡后蒸熟的五色

Soaking and coloring glutinous rice is the most critical step in making Five-color Glutinous Rice. In the process of dyeing, before soaking rice with black, yellow, red and purple dyeing juice, the temperature of the juice should be kept at 38-45℃, and the ratio of juice to glutinous rice should be 1.5 : 1. Next, put the rice into different juices and soak it for about 4-8 hours to color. Put the soaked and dyed glutinous rice into the cask with the bamboo net. Then put it into a big iron pot, add a small amount of water in the pot, steam for about 2-3 hours. The Five-color Glutinous Rice steamed at high temperature is colorful and plump, with the fragrance of plants.

五彩缤纷的糯米 ◆除白色外，其他颜色的糯米均需浸泡上色
Colorful glutinous rice ◆Except for white, all the other colors of glutinous rice need to be soaked and dyed

糯米饭堪称药食同源的天然佳品。

传承：五色糯米饭串起多彩的壮族文化

时光荏苒，时代变迁，五色糯米饭成为壮族民众节庆时必备的传统祭品、日常食品和送礼佳品，被视为幸福、吉祥的象征，体现了壮族人的祭祀文化、稻作文化、节庆文化、饮食文化、礼仪

文化。

壮族家家户户都会做五色糯米饭，据说人们常常以糯米饭的颜色是否鲜艳、味道是否香甜来评定女主人是否为真正的巧妇。在平时的一些喜庆日子，如新居落成、婴儿满月等，壮家人都制作五色糯米饭，以馈赠邻舍、招待亲友。到了每年的“三月三”、清明节等传统节日，五色糯米饭更是家家必做的传统美食。

从2014年起，广西在全国性法定节假日的基础上，将“三月三”传统节日列入自治区的节假日，各地相继举办“三月三”节庆活动，促进文旅产业发展。五色糯米饭作为“三月三”的必备美食，也随着这一节庆影响力的扩大而广为人知。

近年来，政府加大对“非遗”的保护和传承力度，目前已建有16个壮族五色糯米饭制作技艺传承基地，并通过举办“非遗”进校园活动，培养青少年对“非遗”饮食文化和制作技艺的兴趣。

潘红华表示，“非遗”的保护和传承，绝不是简单的“复制、粘贴”，只有通过市场经营和模式创新，“非遗”饮食才能在新时代环境下焕发更大的活力。

以2018年成立的“五彩壮乡”五色糯米饭传承基地为例，其设有壮族古壮字墙、稻作文化展区、居住文化展区、饮食文化展

Almost every family in the Zhuang can cook Five-color Glutinous Rice. It is said that they often evaluate whether the hostess is a brilliant woman based on the color and taste of glutinous rice. On some festive days, such as the completion of a new house and the one-mouth celebration of a baby, the Zhuang will make Five-color Glutinous Rice as a gift to present neighbors and entertain relatives and friends. On traditional festivals such as the third day of the third lunar month and Qingming Festival every year, Five-color Glutinous Rice is a traditional delicacy that every family must make.

区、制作技艺培训与体验展区等多个展区，通过图文、影像、现场体验等方式对壮族五色糯米饭制作技艺进行展示和宣传。参观者可以现场采摘染料植物、学习制作工艺、体验制作过程，在充满壮族元素的空间中，沉浸式地了解壮族特色文化。同时，基地通过政府引导、社会参与、传承人授徒等方式，广泛开展"非遗"展示展演、"非遗"研学等活动。在基地里，由传承人负责提供壮族五色糯米饭制作技艺方面的技术指导，由附近的村民负责稻谷、染料植物的种植。

"五彩壮乡"五色糯米饭传承基地通过"传承基地＋研学开发"和"公司＋合作社"的推动方式，以展示和体验壮族民俗文化为切入点，走出了一条产业引领、文旅融合、多元发展的新路子。

作为"非遗"技艺传承人，潘红华在发扬传统的基础上，不断进行产品研发，使产品从"现做现吃"向"适合存储"，再到如今的"即热即食"转变，以便更好地适应市场需求。同时，注重挖掘创新，按照壮族民俗，开发研制不同时令、不同风俗的"五色"粽子、汤圆、米粉、油茶等系列特色健康产品，并推出祭祖、节庆等精美套装产品，广受市场欢迎。

In recent years, the government has strengthened the protection and development of intangible cultural heritage. At present, there are 16 bases for the promotion of production techniques of Zhuang Five-color Glutinous Rice. In 2018, the "Colorful Zhuang Township" Five-color Glutinous Rice heritage base was officially established, and the Zhuang Five-color Glutinous Rice making skills were promoted and displayed through pictures, images, on-site experiences and other forms.

潘红华表示，未来将继续在五色糯米饭制作技艺上下功夫，争取将它打造成为标准化产品，让每个人都有机会品尝。

以匠人之心传承技艺，以创新之意开发产品，由五色糯米饭串起的多彩的壮族文化，将掀开“五彩广西、美好生活”的时代新篇章。

Through the promotion methods of "heritage base + research and development" and "company + cooperative", the "Colorful Zhuang Township" Five-color Glutinous Rice heritage base has blazed a new path of industry-led, cultural and tourism integration and diversified development.

五色糯米饭　◆不可多得的药食同源的天然佳品

Five-color Glutinous Rice　◆A natural delicacy, rare and dual-purpose for medicinal and edible uses

北海沙蟹汁：天然发酵 咸鲜醇美

Beihai Sand Crab Sauce: naturally fermented, salty and delicious

“非遗”饮食名片

◆ **北海沙蟹汁制作技艺**

类别：传统技艺

级别：自治区级“非遗”代表性项目

批次：第七批

地域：广西北海

俗话说：“靠山吃山，靠海吃海。”沙蟹汁是广西北海特有的一种作料，其以自然生长在沿海滩涂上的沙蟹为主料，通过自然发酵的方法制成。在烹饪的过程中，放入适量的沙蟹汁，即便不放一点盐，也能为食物

扫码看视频

Beihai Sand Crab Sauce is a unique natural dipping sauce in Beihai, Guangxi, which is made from sand crabs growing on coastal beaches and made by natural fermentation. Without a little salt, it can bring salty and sweet taste to food, which is very popular. In 2018, Beihai Sand Crab Sauce Making Technique was selected into the seventh list of representative projects of Intangible Cultural Heritage at the autonomous region level in Guangxi.

带来咸鲜、清甜的口感，因而沙蟹汁深受当地食客喜爱。2014 年经《舌尖上的中国（第二季）》报道之后，沙蟹汁作为北海特产开始广为人知。2018 年，北海沙蟹汁制作技艺入选广西第七批自治区级“非遗”代表性项目名录。

溯源：历史悠久　厚积薄发

北海地处广西南端，北部湾东北岸，是我国最早的对外通商口岸和海上“丝绸之路”的起点之一，历史上是云、贵、川、桂、湘、鄂等省区与海外贸易的主要商品集散地之一。多地文化的碰撞、交融，形成了北海开放包容又极具地域特色的饮食文化。沙蟹汁便是北海饮食文化的代表性符号之一。

北海三面环海，气候条件得天独厚，一年四季盛产蟹类等海产品，为制作沙蟹汁提供了充足的原料。

人类饮食从生食发展到熟食，经历了漫长的过程。沙蟹汁的制作历史悠久，其技艺源自早期渔猎先民腌制生鲜鱼类食物这一技艺。渔民在沙滩上捕捉到沙蟹后，为了最大限度地保持沙蟹的鲜味，采用生鲜腌制的方式，制作出了流传至今、美味鲜香的沙蟹汁。

数百年来，沙蟹汁一直是一种非常小众的作料，仅为北海当地

Surrounded by the sea on three sides, Beihai's unique climatic conditions make it rich in crabs and other seafoods throughout the year, which provides the main materials for making Sand Crab Sauce. The production technique of Sand Crab Sauce can be traced back to the fishing and hunting ancestors who mastered the skill of pickling fresh sea food. After catching sand crabs, in order to keep them fresh and maintain their original flavor, fishermen used pickling methods to make delicious Sand Crab Sauce that has been handed down to this day.

人所熟知。2014 年 5 月，央视热播纪录片《舌尖上的中国（第二季）》在第 6 集《秘境》中介绍了沙蟹汁，虽然只有短短的几分钟时间，但还是让这个极具地方特色的作料迅速蜚声大江南北。

一碗沙蟹汁，浓缩了大海最原始的味道，朴实又不失惊艳，这是海洋赐予人们的味觉享受，也是民间智慧和饮食创造力的集中体现。如今，沙蟹汁已成为北海美食的一张新名片。

技艺：杵碎捣烂　密封发酵

制作沙蟹汁时需将新鲜沙蟹清洗干净后摘除内脏，用木棒捣碎蟹身，加上盐等辅料，最后装罐密封腌制。

北海沙蟹汁制作技艺自治区级代表性传承人周君翠自小就是吃着沙蟹汁风味美食长大的，在她的记忆里，沙蟹汁就如盐或酱油一样，是日常做饭常用的一种调味品。在母亲和奶奶的言传身教下，周君翠学会了做沙蟹汁。在她看来，制作沙蟹汁的关键可概括为原料新鲜和制作精细两点。

（1）抓沙蟹

海水退潮时，北海海滩上时常有大量沙蟹出来活动。这类蟹的大小如拇指，行动极其敏捷。周君翠介绍，抓沙蟹是个经验活，不

For hundreds of years, this condiment with local characteristics has been confined to Beihai market. In May 2014, the popular CCTV documentary *A Bite of China (Season 2)* introduced the Sand Crab Sauce in its sixth episode, *Mystical Land*. Although it was only shown for a few minutes, this condiment with strong regional characteristics quickly gained national fame. Nowadays, Sand Crab Sauce has become a new symbol of Beihai cuisine.

仅要眼明手快，还要懂得看潮位、观气候、算时间。

在北海，渔民们通过“眼子”来计算涨退潮的时间，起潮的第一日为“一眼子”，第二日为“两眼子”，以此类推，15 日为一周期。抓沙蟹的时间一般在晚上，从“三眼子”开始，每一眼子，相差 50 分钟左右，由此计算赶潮时间，才能抓到更多的沙蟹。

此外，气候对抓沙蟹的影响颇大。冬季天气很冷的时候，沙蟹是很难抓的。在其他季节，如果遇下雨天或天气炎热，沙蟹很少出来活动，因此很难抓；逢台风来临前，以及下雨后刚出大太阳时，沙蟹成群结队出来活动，比较容易抓。这些流传下来的经验是世代沿海而居的北海人生活及劳动智慧的总结。

（2）清理沙蟹

抓到的新鲜沙蟹，先在海边用海水进行初步清洗，回到家后还要用干净的淡水反复清洗，直到将沙蟹清洗干净为止。清洗的动作要快，防止沙蟹死亡。然后，挑选鲜活的沙蟹，将其腹底的脐盖掀掉，挤出脐部污物，仔细去除内脏。

（3）杵碎捣烂

把处理好的沙蟹放到洁净、干燥的瓦盘或类似捣臼的器物里，加入盐、姜、辣椒等辅料，用干净的木棒杵碎捣烂。辅料投放比例、

Catching sand crabs needs not only to be quick, but also to have a knowledge of the tide level, tidal time, and climate. After preliminary cleaning the fresh sand crabs caught at the seaside, people come back and wash them repeatedly with clean water. Then it is necessary to lift off the navel cover at the bottom of the abdomen of the thumb-sized sand crabs and carefully remove the viscera.

沙蟹 ◆制作沙蟹汁最重要的原料

Sand crabs ◆The most important ingredient for making Sand Crab Sauce

投放时间和顺序是这个步骤的关键，这属于独家秘籍，决定了沙蟹汁品质和口味的差异。

（4）密封发酵

把制作好的沙蟹汁分装到瓦罐里，并放在阴凉处密封保存，使之自然发酵。微生物产生的蛋白酶，把沙蟹中的蛋白质水解，得到小分子的多肽和游离的氨基酸，更易于人体吸收。发酵是制作沙蟹汁最重要的环节，整个过程不能碰水，不能晒太阳，要保持适宜的温度。周君翠说："发酵不好会导致前面的工艺全部作废，使沙蟹汁产生很重的腥味。"

发酵时间与温度密切相关：冬天发酵时间长，一般在 30 天左右；夏天发酵时间相对短些，大概在 20 天左右。发酵好的沙蟹汁鲜香醇厚，颜色呈茶褐色。

沙蟹汁的制作过程看似简单，但实则步步精细。由于沙蟹汁是发酵物，整个制作过程一定要保持干净卫生。细节上把控不好，如制作工具不够洁净、温度没控制好等，都可能导致发酵失败。可见，做好沙蟹汁并不容易。

沙蟹汁的食用方法多样，烹饪时可用于调味提鲜。炒青菜或者煲汤时放一点沙蟹汁，能增加菜和汤的鲜味，让人吃后回味无穷；

Put the processed sand crabs into a disinfected, cleaned and dried tile tray or other similar mortars. Add salt, ginger, peppers and other accessories, and mash the mixing with a clean stick. After that, put the Sand Crab Sauce into a crock, seal it in a cool place, and let it ferment naturally. The production process of Sand Crab Sauce seems simple, but it needs to be meticulous. For fermentation, the whole production process must be clean. If the details are not well controlled, such as tools are not clean enough or the temperature is not well controlled, the final fermentation will fail.

制作沙蟹汁 ◆千捶百捣凝成鲜

Crafting Sand Crab Sauce ◆Hammered and pounded into an excellent delicacy

就算是最朴实无华的白粥，只要放上一点沙蟹汁，也能让人感受到舌尖最原始的味蕾冲击，一碗寡淡的白粥顿时也有了无尽生气。沙蟹汁焖豆角，正是《舌尖上的中国（第二季）》推荐的北海特色菜，这道菜于2018年被评为广西十大经典名菜之一。其具体做法是：待瓦煲中的豆角半熟时，放入几勺沙蟹汁拌匀，盖上盖子继续加热，待熟时揭开锅，豆角的豆香气混着沙蟹汁的海鲜味，鲜香扑鼻，让人胃口大开。

沙蟹汁还可以作为食物蘸料。北海人吃白切鸡、白灼虾、白切猪肉等食物时，喜欢蘸沙蟹汁吃，以提味增鲜、消食解腻。

沙蟹汁含有丰富的蛋白质，大量的钙、磷、钾、钠、镁、硒等微量元素，对身体有很好的滋补作用。在北海，人们常用沙蟹汁与其他食材相搭配，烹制养生菜肴。

传承：延续传统　谱写新篇

周君翠早年在北海经营了一家渔家乐，其主打产品沙蟹汁风味美食大受食客欢迎，她便萌生了把北海沙蟹汁推广出去的想法。然而，沙蟹汁多以家庭手工制作为主，难以大批量生产以满足市场需求。2013年，周君翠开始筹建沙蟹汁工厂，期望通过扩大生产规模，

Sand Crab Sauce is mainly used for seasoning and freshness when cooking. When stir-frying vegetables or making soup, add a little Sand Crab Sauce can enhance the flavor. When making boiled chicken slices, boiled prawns, it will be perfect to use Sand Crab Sauce instead of scallion sauce. In her early years, Zhou Juncui came up with the idea of promoting Beihai Sand Crab Sauce. In 2013, Zhou Juncui began to build a Sand Crab Sauce factory. In 2014, Zhou Juncui and her family helped shoot the content of "Beihai Sand Crab Sauce" in *A Bite of China 2*, which caused great response as soon as it was broadcast. This magical seasoning opened its new marketization.

让这道具有地方特色的作料为更多人所知。

2014 年，周君翠一家协助拍摄了《舌尖上的中国（第二季）》中有关沙蟹汁的内容，节目一经播出便引起了巨大反响，这道神奇的作料开始被更多人所知，也开启了其市场化的新历程。某电商平台的官方数据显示，在节目播出后的一周内，沙蟹汁的搜索量增长了将近 500 倍，购买量也得到了大幅度提升，成了一炮走红的明星产品。

借此契机，周君翠加快了企业化经营的步伐。为了保障产品的

沙蟹汁焖豆角　◆广西十大经典名菜之一

Braised String Beans with Sand Crab Sauce　◆One of the ten classic dishes of Guangxi

稳定性，周君翠在传统制作工艺的基础上，通过不断试验，制定了第一份有关沙蟹汁生产的企业标准，该标准涉及产品菌落含量、香度、鲜度、颜色与发酵时间等各个环节，让以往仅凭个人经验把控的产品实现标准化生产。周君翠还通过线下特产渠道和线上网络平台，将这一地方特产推向全国。

随着沙蟹汁在全国知名度的提高，一些问题也开始显现，最突出的是地域口味问题。因为沙蟹汁的整个制作过程没有加热煮熟的步骤，所以其带有一种特别的味道，当地人习以为常便觉得它美味无比，而很多外地人不习惯则难以接受这种味道，即使慕名尝鲜，浅尝后也敬而远之。

为了让外地人能更好地接受沙蟹汁，周君翠不断研究新菜式，完善沙蟹汁产品的食用说明书，通过美食制作展示沙蟹汁的食用场景，开拓沙蟹汁的食用价值。另外，她还在游客集聚的北海老街设立了体验店，让更多的游客了解沙蟹汁、品味沙蟹汁。周君翠持续投入资金加大新品研发力度，目前已经开发了四五种衍生产品，使沙蟹汁从地方特色作料向日常作料转变，从而迈向更广阔的市场。

在发展北海沙蟹汁的道路上，周君翠认为最重要的一点还是生产者要完善制作技艺："不管其他方面做得多么好，质量才是根本

As Beihai Sand Crab Sauce has gradually become famous in the national market, some problems were beginning to emerge, of which the most obvious one is personal taste. In order to enable outsiders to better accept Sand Crab Sauce, Zhou Juncui constantly studied new dishes and refined instructions for eating Sand Crab Sauce, demonstrated the eating methods of Sand Crab Sauce through delicacy cooking, and explored the edible value of Sand Crab Sauce. Besides, she continued to invest in development of new products, which expanded the market of Sand Crab Sauce from local condiments to daily condiments.

之道。”

沙蟹汁的兴起也吸引了更多关注。在有关部门的组织和协助下，北海沙蟹汁生产企业积极参加各类“非遗”展示活动，以会宣传，以展促销，让更多人了解沙蟹汁这个极具地方特色的作料。2021 年，沙蟹汁在美食纪录片《开动吧！海鲜》的第 5 集《沙蟹汁——原来你是“酱”的螃蟹》中播出，再一次成功“出圈”，受到了广泛关注。

海天一色蓝，潮涌北部湾。北海沙蟹汁，这个“亿万年秘境里酿造出的奇迹”，正穿浪而行、倍道而进，谱写着新的秘境篇章。

The rise of Sand Crab Sauce has also attracted more attention. With the assistance of government department, Beihai Sand Crab Sauce production enterprises actively participated in various intangible cultural heritage exhibition activities both inside and outside Guangxi, and increased publicity and promotion efforts to let more people know about this unique cuisine of Beihai. In 2021, the fifth episode "Sand Crab Sauce" of the seafood documentary "Let's Go! Sea Food" was broadcast, and Beihai Sand Crab Sauce once again became well-known.

舌尖上的
广西『非遗』

食之文化：『那』边风味独好

Food Culture: "Na" has a unique flavor

"非遗"饮食是广西饮食文化的重要组成部分，其不仅是口腹之欲，还包含着历史文化的传承、风俗民情的展现、游子乡愁的寄托及地域身份的认同等。透过螺蛳粉，人们可以看到早在2万多年前，柳州地区"白莲洞人"已懂得采集、加工和食用螺蛳；越过五色糯米饭，可以看到壮家人在虔诚祭拜龙母和祖先，祈望着风调雨顺、五谷丰登；通过桂林米粉，可以看到白先勇先生及如他那般远离故土的游子在喃喃低语，述说浓浓乡愁；穿过六堡茶的氤氲茶香，可以看到无数背井离乡的劳工在翘首遥望着故土……

穿越"非遗"饮食的历史风雨，人们看到了广西源远流长的稻作文化，以及由此衍生出的精彩纷呈的饮食文明；看到了几千年来生活在八桂大地上的人辛苦劳作所创造的既各具特色又相互交融的饮食风俗；看到了来路，展望着归途。

Intangible cultural heritage food is an important part of Guangxi's food culture. Through the historical ups and downs of intangible cultural heritage food, people see the long-standing rice culture of Guangxi and the wonderful food civilization derived from it, and see the food customs created by the hard work of the people who have lived on the land of Guangxi for thousands of years, which are both unique and intertwined.

稻作根基：赖“那”而食

The foundation of rice cultivation: depending on "Na" for food

『那』在壮语中意为『稻田』。以稻食为基础、赖『那』而食，是广西饮食最显著的特征。

稻作文化是指人们以水稻种植为主要生存和发展方式的文化，由此而衍生的有关衣食住行的种种风俗都属于稻作文化的范畴。

早在2012年，中国科学院国家基因研究中心课题组以广西南宁市隆安县等地的野生稻为蓝本，通过基因组重测序和序列变异鉴定，从全球不同生态区域中选取400多份普通野生稻，绘制出水稻全基因组遗传变异图谱，提出了“分布于广西的普通野生稻与栽培稻的亲缘关系最近”的研究成果，表明广西是水稻的驯化起源地。

经过群体遗传学分析，大致可推断出栽培水稻的扩

As early as 2012, the research group of the National Genetic Research Center of the Chinese Academy of Sciences took wild rice in Long'an County as a model, through genome resequencing and sequence variant identification, drew a map of rice genome-wide genetic variation from more than 400 common wild rice samples selected from different ecological regions around the world, and proposed the research results of "the closest kinship between common wild rice distributed in Guangxi and cultivated rice", indicating that Guangxi is the origin of rice domestication.

散路径：人类的祖先首先在广西境内的珠江流域利用当地的野生稻种，经过漫长的人工选择驯化出了粳稻，随后粳稻往北逐渐扩散；而往南扩散的则进入东南亚，在当地与野生稻种杂交，产生了籼稻。这一重大的研究成果，发表在了世界权威学术杂志《自然》上。

2015年，研究人员在隆安县娅怀洞遗址中发现了距今约1.6万年的稻类遗存。这是我国目前发现的人类利用野生稻最早的考古证据，证明了壮族先民在这里繁衍生息，古老的稻作文化由此发轫。

“那”在壮语中，意为“稻田”。壮族先民世代以稻作维系生存，他们开垦稻田，生产水稻。广西自古就是长江以南稻米的重要产区之一，素有“北有五常稻花香，南有广西晚百香”的美誉。东汉杨孚的《异物志》有“稻，一岁夏冬再种，出交趾”的记载，说明再种稻在当地种植普遍。西晋时期的文人郭义恭在《广志》一书中所说的“南方有蝉鸣稻，七月熟”，说的便是上林的水稻。以稻食为基础、赖“那”而食，是广西饮食最显著的特征。

大米是广西人的主食。广西属于亚热带季风气候区，气候炎热，夏季漫长，人们极易食欲不振，因此相比米饭，广西人更爱大米制成的粉和粥。

粥，在古代称为“糜”，喝粥的现象从先秦就开始出现，流传

In 2015, researchers discovered rice remains dating back more than 16,000 years in the Yahuai Cave site in Long'an County. This discovery represents the earliest archaeological evidence of human use of wild rice found in China, which indicates that the ancestors of the Zhuang people flourished in this region, giving birth to an ancient rice culture. The Zhuang language's term "Na" translates to "rice field". Therefore, the prominent feature of Guangxi's diet is its heavy reliance on rice and the cultivation practices associated with "Na".

千年，明代李时珍在《本草纲目》中列有多种粥的做法和药效，可见在那时粥就是公认的养生保健佳品了。食粥是广西人日常饮食中不可分割的一部分，尤其是夏日，一碗好粥，既爽口又能增进食欲，最能安抚人心。除了常喝的白粥，广西人将粥制作成各种花样，有玉米粥、八宝粥、绿豆粥、小米粥、生料粥、海鲜粥、酸粥及青菜瘦肉粥等。大量调查资料表明，巴马百岁老人中有 84% 的人每天早餐都要喝粥，最常吃的主食是玉米粥——把当地出产的玉米打碎或磨成粉，加入地瓜或者南瓜一起煮，巴马人把这种粥叫作“百岁饭”。在广西很多地方，还经常可以看到自助粥摊，在网上被戏称为“广西人才懂的六元自助餐”，由此可见粥在广西人日常饮食中的重要性。

米粉也是广西人不可缺少的主食。广西米粉的制作方式很多，从无汤干捞到浓汤重味，从生榨现煮到炒粉卷粉，制法之繁多，令人叹为观止。粉的种类也极为丰富，有重口味的螺蛳粉、桂林米粉、老友粉、宾阳酸粉、生榨米粉和红油米粉等，有清淡鲜美的生料粉、海鲜粉、牛巴粉、烧鸭粉与鸡肉粉等，还有制作工艺独特的粉虫、粉利及卷筒粉等。从大都市到小乡镇，从高端奢华的五星级酒店到偏于一隅的街边小摊，都少不了米粉的身影。无论你是来自天南还

Guangxi has a hot climate, so Guangxi people prefer rice noodles and porridge made of rice. In addition to the simple congee that is often drunk, Guangxi people make congee into various patterns. In addition, rice noodles are also an indispensable staple food for Guangxi people, from mixing the rice noodles with marinade to thick soup with heavy taste, from raw squeezed and freshly boiled to fried and roll, the variety of recipes is amazing.

是地北，在广西总能找到适合自己口味的一碗粉。

除了“嗜粥”“嗜粉”之外，广西人还用稻米制成了品类繁多、独具特色的稻米制品。壮族人习惯用糯米制作各种节日食品，如五色糯米饭、粽子、年糕、糍粑、米糕、汤圆及油团等，其中以五色糯米饭最具代表性。每年“三月三”等节日，壮族人家家户户都会制作五色糯米饭，用天然植物提取的汁液将糯米染成不同颜色，用其祭拜祖先，馈赠亲友。此外，广西还有各种米制小吃，如米饼、沙糕、米花糖、云片糕和糖环等，这些小吃味道甜美，口感香糯，极具地域特色。

北海、钦州和防城港一带则流行着粄类小吃，其用粳米、麦粒等作为主料，加入各种辅料制作而成。粄的种类繁多，口味多样，如水粄、盖粄、大笼粄、艾粄、叶麻粄、梢叶粄、白糍粄、发粄及粟米粄等，因此当地有“满汉全席，不如吃粄”的说法。

在广西，米酒的制作历史悠久。据史料记载，早在汉朝时期，广西的酿酒业就已经兴盛起来，以苍梧所产的酒最为知名，这是一种以大米为原料、以传统发酵方法酿造的米酒。及至如今，广西大多数知名的酒类品牌，如“三花酒”“湘山酒”“天龙泉”“东园家酒”等，依然是米香型白酒。

In addition to loving congee and rice noodles, Guangxi people have further expanded the food chain of rice and made a variety of unique rice products. The Zhuang people are used to using glutinous rice to make various festive foods, such as glutinous rice, zongzi, New Year cake, rice dumplings, tangyuan and oil balls, etc. In addition, a variety of rice snacks, such as rice cakes, sand cakes, rice candy, square-shaped white glutinous rice-cake, sugar rings, etc., the taste is sweet, fragrant and glutinous, with strong regional characteristics.

此外，以米入菜，亦是广西的饮食特色。例如，有用盐和蒸熟的糯米等腌制的酸鱼，有将糯米、肉末和香菇等塞进油豆腐、苦瓜圈等做成的各种酿（豆腐酿、苦瓜酿），有将糯米与新鲜猪血等相拌和后塞进洗干净的猪肠中制成的血肠，有用拌有猪油、白糖的糯米饭与 8 种干果、蜜饯相混合制成的八宝饭……广西人将糯米与各种食材，以独具匠心的制作方法融合在一起，形成别具风味的地方饮食。

几千年来，广西人据“那”而作，凭“那”而居，赖“那”而食，靠“那”而穿，依“那”而乐。依托在“那”文化根基上的广西饮食，如生命力强盛的榕树一样，在八桂大地上茁壮成长，形成了独具特色、自成一派的饮食文化。

In Guangxi, rice wine enjoys a long history and widespread popularity. Historical records indicate a thriving winemaking tradition in the region dating back to the Han Dynasty. Today, many of Guangxi's most renowned wine brands continue to produce rice wine. In addition, rice is also a characteristic of Guangxi's cuisine. For thousands of years, Guangxi food, which is based on the foundation of "Na" culture, has grown vigorously in this land like a banyan tree with strong vitality, forming a food culture of its own.

食材丰富：山珍水产皆可入菜

Abundance of ingredients: all kinds of delicacies and aquatic products can be used in dishes

山珍水产、家禽蔬果皆可入菜，造就了广西蔚为大观的饮食菜系。

广西属于亚热带季风气候区，光照充足，雨水充沛。广西的地形地貌复杂，土地类型多样，动植物资源丰富。南部濒临的北部湾，不仅是中国著名的渔场，也是世界海洋生物物种资源的宝库。

广西优越的山水环境孕育了数不胜数的各类食材。广西作为全国六大稻作区之一，是南方籼米的主要产区，有着“八桂粮仓”之称；广西是家禽生产和消费重点区域，家禽地方品种数量全国第一；广西是全国重要的“南菜北运”蔬菜基地、全国最大的秋冬菜基地；广西是中国食糖的主产区，糖料甘蔗的种植面积和食糖产量占全国 60% 以上，据说，全国每三勺白糖中，

Guangxi belongs to the subtropical monsoon climate zone, enjoying ample sunshine and rainfall. Guangxi boasts a complex terrain and landforms, diverse land types, as well as rich animal and plant resources. The Beibu Gulf, bordered in the south, is not only a famous fishing ground in China, but also a treasure trove of marine species resources in the world. Numerous mountain ranges and water sources have bred countless kinds of ingredients.

就有两勺是来自广西；广西亦是中国水果大省，水果丰富，具有稀、优、大、全等特点，其中柑橘、香蕉、火龙果、百香果、杧果和柿子等水果的产量排在全国第一，是名副其实的“水果王国”；广西拥有丰富的富硒农产品资源。

“江作青罗带，山如碧玉篸”，优越的地理环境使得广西物产丰富，稀缺的好山好水成就了优质的原生态食材。

丰富、优质的食材是广西美食的保障，山珍水产、家禽蔬果皆可入菜，造就了广西蔚为大观的饮食菜系。尤其是近年来，饮食类节目的风行、螺蛳粉的火爆，使广西美食受到了很大的关注。陈晓卿导演就对广西美食颇为称赞：“广西是中国饮食最复杂的地方”；“广西美食最大的特点就是多样，每走 20 公里就有新的惊喜”。他还专门拍摄过一部名为《秘境广西》的纪录片，用三集内容才算概括了广西博大精深的饮食文化，引发了众多美食达人及媒体对于广西美食的关注。

在这场美食狂欢中，广西一些地域化、小众化的饮

Abundant and high-quality ingredients are the guarantee of Guangxi's deliciousness, and mountain and aquatic products, poultry, vegetables and fruits can all be used in the dishes, creating its magnificent cuisine. Especially in recent years, the popularity of food programs and the favorite of Luosifen have made Guangxi cuisine receive great attention and excavation. Director Chen Xiaoqing shot a documentary called "Secret Guangxi", which used three episodes to summarize Guangxi's broad and profound food culture.

食，如炸昆虫、羊瘪汤和牛欢喜等，也得到了一定程度的传播，广西人被调侃为“食物链顶端的王者”。这展现了广西饮食的多样性与广泛性。

广西可入菜的食材丰富，广西饮食注重食材品质，天然生态、原汁原味是其主要特征。比如，制作广西的“草根美食”武鸣柠檬鸭时一定要选用当地散养的鸭以及当地出产的皮薄多汁的小柠檬；制作全州醋血鸭时要选用全州文桥土麻鸭；梧州纸包鸡制作时要优选正宗的当地农家山地散养的三黄鸡；制作合浦大月饼时要选用当地产的黑毛土猪，并加入合浦特产鸡头荔枝蜜、土酒和五香粉等调料腌制；制作荔浦芋扣肉时一定要选用荔浦芋……注重食材，精选食材，追求天然，崇尚生态，是广西饮食的一大特点。

美食家左壮在《美食之心：味蕾之巅的料理智慧与心灵对话》一书中说，他觉得食物越少加工越好，能生吃的就生吃，不能生吃的就用最简单的烹饪方式把最美的味道呈现出来。这也正是广西饮食的加工理念，讲究本味和原汁原味。例如，追求鲜、活、嫩的横县鱼生、北海沙蟹汁，萃取大自然的花、草、树叶和果实等染色制成的壮族五色糯米饭，清淡爽口、猪鸡鸭鹅皆可白切的博白白切菜，等等。即使要调味，也采用天然调味品，如用几十种草药、香料熬

Although the dishes feature a wide range of ingredients, the quality of the ingredients in Guangxi cuisine is never compromised, with natural ecology and original taste being its main characteristics. The food processing concept of Guangxi diet pays attention to the original taste. Even if it is to be seasoned and processed, Guangxi people pay attention to natural flavors, and bring the most authentic taste of food through the combination and blending of ingredients.

制桂林米粉的卤水，用石螺、猪筒骨和各种香料熬制螺蛳粉的螺蛳汤……通过食材间的搭配与调和，激发出食物最本真的味道。

天然、生态的食材，原汁原味的加工手法，成就了一道道广西特色美食。千百年来，在自然山水间，在一蔬一饭中，在时间的打磨下，这些美食一代代延续着。其在大自然的慷慨馈赠下，在食材与风味的相融中特立独行。

Natural and ecological authentic ingredients and original food processing techniques have made Guangxi's special cuisine. For thousands of years, generation after generation has continued to enjoy the most generous gift from nature, and as a result, Guangxi cuisine stands out in its unique combination of ingredients and flavors.

一“味”通行：万物皆可酸

One "taste" for all: everything can be sour

广西人“嗜酸”。在广西，生津开胃的酸是不可或缺的一道口味。

广西饮食的多样性和复杂性不用再赘述，其地域口味被总结为“东甜西酸、南鲜北辣”。但如果有一个味道可以覆盖南北、连通东西的话，那就非酸味莫属。广西的酸多指通过发酵技艺所制作的酸食，其工艺之娴熟，内容之丰富，简直达到了炉火纯青的地步。

在南宁，大街小巷最常见的一种小吃叫“酸嘢”，“酸”是口味，“嘢”在当地话里是“东西”的意思，“酸嘢”意为用果蔬腌制而成的一种酸品小吃。不只萝卜、甘蓝、黄瓜、莲藕、芥菜、生姜和大蒜等菜蔬可以腌酸，菠萝、杧果、木瓜、扁桃、阳桃与李子等水果更是“酸嘢”摊上的“常客”，摊上一般配有辣椒酸醋水、

There is no need to repeat the diversity and complexity of Guangxi's diet, and its regional taste is summarized as "sweet in the east, sour in the west, fresh in the south, and spicy in the north". But if there is one flavor that can cover the north and south and connect east and west, it must be sour. Guangxi's pickle is made by fermentation technology, and its craftsmanship and rich ingredients have reached the perfect level.

椒盐等，供食客随意蘸用。初次来南宁的游客一般看到这种酸酸辣辣的果蔬会觉得不可思议，但当地人却爱之如命，尤其在高温炎热的夏天，一口吃下去，酸辣爽口，故而在当地有着“英雄难过美人关，美人难过‘酸嘢’摊”之说。

广西的酸食不仅普遍，还因地域风俗不同，制作工艺有异，口感味道也各具特色。例如，南宁的“酸嘢”是用酿造食醋做成的，把新鲜的果蔬切成片状或块状，然后用酸醋浸没入味，以酸、脆、爽的口感著称；

用新鲜杧果制成的“酸嘢” ◆南宁大街小巷常见的一种酸品小吃
The pickles made of fresh mangoes ◆A common sour snack seen on the streets and alleys of Nanning

都安酸的精髓在于红糖醋，加入了红糖醋的酸制品酸甜爽口；玉林酸分为两种——鲜果酸和泡制酸，鲜果酸加了秘制的黄糖调味，口感浓郁鲜甜，泡制酸则是将果蔬泡在秘制的酸水中，直至完全入味，这种酸捞出后不用加调料都很好吃；钦州酸是将当季果蔬用醋、糖和辣椒等进行腌制，再加上特制拌料——甘草盐或甘草水，其不仅能中和果蔬里的酸味，还能增添风味，让人吃后回味悠长。

除了这种可以直接吃的酸品小吃，还有炒菜或者配菜用的酸料，如酸笋、酸柠檬、酸芥菜、酸豆角和酸辣椒等。可以说，酸在广西饮食里几乎无处不在，家家户户都有腌酸的老坛子或玻璃缸，一年四季都用它来腌制不同的当季蔬菜。腌好的酸料可以配粥吃，也可以入菜。例如，因《舌尖上的中国》这部纪录片而出名的酸笋，就是广西菜里常见的食材，酸笋炒牛肉、酸笋炒田螺、酸笋黄豆焖鱼仔、酸笋炒肥肠、酸笋炒鸡、酸笋焖鸭、酸笋炒花甲、酸笋炒蕹菜梗……只有想不到的，没有炒不了的。坛子里的酸水被当地人称为“坛子醋”“坛酸醋”“老坛醋水”等，也是炒菜的最佳作料。例如，制作全州醋血鸭时放入的醋血，就是杀鸭取血后，把血滴入酸水之中，迅速搅拌混合均匀而成的，这是全州醋血鸭的“神来之笔”，有了这些酸水，鸭血即使放两三天也不会凝成块状，还可以去腥

In Nanning, one of the most common snacks on the streets and alleys is called "pickle". They are made of fruits and vegetables. Guangxi's pickle is not only common, but also has different production techniques due to different regional customs, and has different taste.

生津开胃的酸笋 ◆广西菜常用的一种食材

Sour bamboo shoots that stimulate saliva production and appetite ◆A common ingredient in Guangxi cuisine

提鲜。

在广西，酸料也是米粉和粥的“绝佳拍档”。广西的各种米粉经常与酸料搭配，如煮老友粉时要放酸笋。广西米粉店的作料台上放着各种自助添加的酸料，最常见的有酸豆角、酸笋、酸辣椒和酸柠檬等。自助粥铺里的酸料更是种类丰富，让人眼花缭乱，酸料可

In addition to this "sour snack" that can be eaten directly, there are also sour ingredients for stir-frying or side dishes, such as sour bamboo shoots, sour lemons, dried sour radish, sour beans, sour peppers, etc. Acid is almost ubiquitous in the Guangxi's diet, and every household has old jars or glass jars with pickles. The pickles can be eaten with congee and can be added to dishes, and they are also "excellent partners" for rice noodles. Squeezed Rice Noodles, sour congee, sour fish, etc., which are naturally fermented by a unique process, show the dietary customs of Guangxi people who love to eat sour food.

自助选取，搭配白粥食用。

此外，用独特工艺天然发酵而成的生榨米粉、酸粥、酸鱼等，更显示了广西人“嗜酸”的饮食习俗。生榨米粉是将大米发酵后，经过多个工序加工煮制而成的一种发酵型米粉，其以软、滑、香的口感，以及发酵后乳酸菌赋予的与众不同的“酸馊味”而闻名。在龙胜侗寨，可以说是“无菜不腌、无菜不酸”，各种肉类用盐和蒸熟的糯米等腌制发酵，逐渐水解转化产生的乳酸，时间愈久，味道就越为醇厚。让陈晓卿导演念念不忘的侗族酸鸭子，“又酸又臭，却香艳无比”。

酸之于广西，是绵延在日常中的饮食底色，究其原因，与广西的地理位置、气候与习俗等息息相关。广西总体是山地丘陵性盆地地貌，四周都是绵延不断的山岭。广西的少数民族大多依山而居。先民的吃食一般都是自产自销，受季节时令影响较大，为了不使食物变质腐烂，聪明的人们把食物制成酸食。只要制作中把坛子的密封工作做到位，酸食可以保存很久，而且还带有可替代咸味的复杂味道。根据记载，这种“以酸代盐”的食物处理方法可以追溯到2000多年前。

广西高温闷热的气候，也是酸食风靡的原因之一。在高温多

Acid in Guangxi is the base color of the diet that extends in daily life, and the reason is closely related to Guangxi's geographical location, climate, and customs. Guangxi is mainly hilly and mountainous, and the food of the ancestors was generally self-produced and self-sold. In order not to spoil the harvested food, smart people made the food into sour food. The high temperature and sultry climate is also one of the reasons for the popularity of acid food.

万物皆可酸 ◆没有什么食材是广西人不能用来做成酸食的

Everything can be turned into pickles ◆There's nothing that Guangxi people can't make into sour dishes

雨的夏季，人们像是被困在蒸笼里闷蒸着，极易吃不下饭。在五脏与五味的关系中，酸味入肝，具有收敛之性，可生津解渴、解腻开胃、促进消化、增进食欲，久而久之，当地人慢慢形成爱吃酸的习俗。清嘉庆年间白山司（辖境主要在今广西马山县）巡检王言纪写有《白山司志》，记录了当地人制作酸笋之事，曾提到“椒以祛寒，笋以治热，二物备而水土不灾”。可见，食酸以治热

In Guangxi, a multi-ethnic, multi-cultural and multi-geographical environment, only the sour flavor, crosses ancient and modern times, and penetrates into the marrow of Guangxi people regardless of region, carrying the life wisdom of the Luo Yue ancestors and continuing the unique "Guangxi Taste".

由来已久。

在多民族、多文化、多地理环境的广西，唯有酸之一味，不分地域，跨越古今，沁入广西人的骨髓之中，承载着骆越先民的生活智慧，延续着独具一格的“广西味道”。

酸藠头与酸萝卜 ◆既可直接吃又可用来炒菜的“酸嘢”

Sour pickled garlic bulbs and sour pickled radishes ◆The pickles that can be eaten directly or used for cooking

药食同源：餐桌上的养生

Food and medicine are of the same source: health preservation at the dinner table

『药补不如食补』，将广西的道地药材与食材相融合烹制出来的美食，是在美味和健康间取得的一种平衡。

我国自古就有药食同源的饮食理念和传统，《黄帝内经》曰：“药以祛之，食以随之”；“五谷为养，五果为助，五畜为益，五菜为充，气味合而服之，以补精益气”。《神农本草经》载药365种，其中大枣、人参、枸杞、五味子、地黄、薏苡仁、茯苓、沙参、生姜和葱白等，都是具有药性的食物。明代李时珍编写《本草纲目》，仅谷、菜、果三部就收录了药用食物300多种……丰富的药食理论和实践，让“药补不如食补”的理念深入人心，并形成了蔚为壮观的药膳文化。

广西地处中国的西南边陲，古时人烟稀少，重峦叠嶂，多雨潮湿，瘴疠横行。恶劣的自然环境不仅磨砺出

China has always had a dietary philosophy and tradition of homology between medicine and food since ancient times. The rich theory and practice of medicinal food have made the concept of "medicine supplement is not as good as food supplement" deeply rooted in the hearts of the people, and formed a spectacular medicinal food culture.

广西人勤劳坚韧的品格，还使得广西人就地取材，总结出了食疗养生的饮食文化。传说，桂林米粉卤水的雏形就是将传统中草药与当地香料等一起煎熬制成的防疫药汤；恭城油茶的起源也与之类似，聪明智慧的瑶族人为了对抗恶劣的自然环境，用茶叶、生姜、蒜头等煮水而食，代代相传，逐渐演变成今天人们爱喝的恭城油茶；梧州龟苓膏起源于当地人为了抵抗湿热之症，熬制中草药用于清热解毒的传统；梧州六堡茶最初亦是以药的形式出现的，当地人用来治疗痢疾、祛湿避瘴，其作为“保命茶”在东南亚劳工中风行一时，民间一直有着“三年药，五年宝，十年丹”之说。

除了这些耳熟能详的“非遗”饮食，广西人的日常餐饮中，亦是随处可见食疗养生的习俗。遍布广西很多城市的凉茶铺，便是其中之一，纪录片《本草中国》中曾如此介绍：“凉茶既不凉，也不是传统意义上的茶，它以民间常用复方或单味土产草药煎熬而成。所谓‘上医治未病’，医术的至高境界在于防患于未然，而凉茶就是其中的典型。”由此可知，凉茶本质上就是用中草药熬制而成、具有一定保健作用的药汤，民间有着“饮一杯凉茶，不用找医家”的说法。

梧州凉茶是岭南地区凉茶的典型代表，梧州凉茶配制技艺是以

Guangxi is located in the southwestern border of China, sparsely populated in ancient times, with mountains, rain and humidity, and miasma. It makes Guangxi people take local materials and summarize the food culture of food therapy. The origin legend of Guilin Rice Noodles marinade is that traditional Chinese herbs are blended together with local spices to make epidemic prevention medicine soup; Wuzhou Liupao Tea also originally appeared in the form of medicine.

中医养生理论为指导，结合梧州本土的气候、水土特性，将梧州多种具有药用疗效的道地药材，按特定药方配比后煎水，经净制、切制、称量、浸泡、入煲、加水、熬制和存贮等多道工序熬制成凉茶的一种传统技艺，传承至今已有 1700 多年历史。

钦州凉茶种类众多，按口味可分为苦茶、甜茶两种。苦茶就是以辛、苦、寒、凉的中药为主的凉茶，如癍痧凉茶等；甜茶是以清润甘甜的药材为主的凉茶，如竹蔗茅根水、罗汉果五花茶等。

总而言之，广西的凉茶所用的草药原料大多数是当地及周边地区土生土长的野生品种，如雷公根、茅根、一点红、淡竹叶、鱼腥草、鸡骨草、田基黄、荷叶、金银花、罗汉果、野菊花、地桃花、白纸扇和救必应（铁冬青）等。凉茶配方既有单方，也有复方，以复方为主。凉茶除了能生津止渴，提神醒脑，还对某些疾病有“茶到病除”的作用。

对于很多老一辈的广西人来说，没有凉茶，整个夏天就缺少了些许滋味。而对于年轻一代来说，凉茶文化的影响则在慢慢消退。相比较于传统凉茶，他们更喜欢喝奶茶、咖啡以及商店里售卖的各类改造过的清甜的凉茶类饮料。“良药苦口”的传统凉茶铺越来越少，这从一个侧面反映出新式饮食文化对传统饮食文化的冲击。

In the daily diet of Guangxi people, the customs of food therapy and health care can be seen everywhere. Herbal tea stalls in many cities of Guangxi are one of them. Herbal tea is essentially a medicinal soup with certain health effects made from Chinese herbal medicines, and is a traditional refreshing drink in Guangxi's summer. Wuzhou herbal tea is a typical representative of herbal tea in Lingnan region. There are many types of Qinzhou herbal tea, which can be divided into two types: bitter tea and sweet tea according to taste.

近年来，广西积极推进传统凉茶文化保护工作，如钦州杨氏老牌凉茶制作技艺、梧州凉茶配制技艺分别入选钦州市第七批、梧州市第六批市级“非遗”代表性项目名录等就是这方面的积极行动。传统凉茶不仅传承着文化，也传递着长辈对小辈的关怀，就如年长的人所说：“当你懂得自己到街上寻觅儿时被父母逼着喝的那杯苦凉茶时，说明你已经长大了。”

一杯凉茶润人心，一碗炖品暖人心。广西很多城市大街小巷都有的滋补炖品，也是当地药膳文化的体现之一。广西的炖品各式各样，原汁原味，具有香、甜、补的特点，大多具有扶正祛邪、调养保健的功效。广西的炖品一般是根据经验或药膳食谱配方，将药材、食材和调味品放入瓦盅里，加入清水，小火慢炖而成的。鸽子可滋补益气、祛风解毒，莲藕可健脾开胃、清热凉血和益血补心，山药可补脾益肺、滋肾阴、敛虚汗……对于广西人来说，无数的道地药材和食材在瓦盅中融合成的美食，是在美味与健康间取得的平衡。

广西药材资源丰富，道地药材很多，有“天然药库”“生物资源基因库”“中药材之乡”的美名。广西的中药材品种有 7000 多种，其中道地药材以“桂十味”最为有名，即肉桂（含桂枝）、罗汉果、

The nourishing stews that are found in the streets and alleys of many cities in Guangxi are also one of the embodiments of the local medicinal food culture. Guangxi's stews are varied, original, fragrant, sweet and tonic, most of them have the effects of strengthening the body's resistance, dispelling evil and nourishing health benefits. Generally, according to experience or medicinal dietary recipes, put medicinal herbs, food and condiments into a clay pot, add water, and simmer over low heat. The fusion of countless authentic herbs and ingredients in a clay cup is a balance between delicious and healthy.

凉茶 ◆用中草药熬制而成的药汤
Herbal tea ◆A medicinal decoction made from traditional Chinese herbs

八角、广西莪术（含桂郁金）、龙眼肉（桂圆）、山豆根、鸡血藤、鸡骨草、两面针和广地龙。广西还有“中国天然植物香料库”的美誉。据统计，广西拥有香料植物 270 多种，其中八角、肉桂、茉莉花及金花茶等优势香料稳定占据全球过半乃至七八成的产量。

天然、丰富的药材和香料资源，让广西饮食在药膳、养生、健康领域拥有得天独厚的优势。随着人们养生意识的逐渐增强，药膳文化正逐步成为新潮流，这有利于广西药食同源饮食文化的传承和开发，将大幅提升广西饮食的知名度和影响力。

With rich natural medicinal materials and spices, Guangxi's diet has rich practice and long history in the fields of medicinal diet, health preservation and health. With the enhancement of people's awareness of health preservation, medicinal food culture is gradually becoming a new trend, and the preservation and development of Guangxi's medicine homologous food culture will enhance the popularity and influence of Guangxi's diet.

“广西味道”：舌尖上的文化自信

"Guangxi Taste": cultural confidence on the tip of the tongue

美食带来的不仅是味觉享受，更是对地方文化的触碰与感知。螺蛳粉就是广西对外展示的重要窗口之一。

《礼记》有云“夫礼之初，始诸饮食”，认为饮食活动中的行为规范是礼制的发端。司马迁的《史记》中有“王者以民人为天，而民人以食为天”的记载，将饮食视为国祚稳固的根本。《孟子》曰“食色，性也”，认为饮食乃人之本性。《黄帝内经》云“五味之美，不可胜极”，讲的是五味调和可以带给人美好的享受……饮食文化，是中华文明画卷上的一颗璀璨明珠，在一餐一饭、一饮一啄间，演绎着人间百味，传承着风土习俗。

广西饮食文化作为其中的一个分支，如滚滚的西江之水，在崇山峻岭间奔腾，自成一系，又在平原低缓处汇入，相互融合。其以独特性、丰富性著称于世，又以

Food culture is a brilliant pearl on the picture scroll of Chinese civilization, in every meal, every drink, interpreting the various flavors and keeping local customs. Guangxi's food culture is a branch of China's food culture. It is known for its uniqueness and richness, and accepting all styles with its inclusiveness and peace.

包容性、平和性接纳万方。

近年来，“广西味道”因其神秘、丰富的特色受到了越来越多的关注。其以“非遗”饮食为代表，成为广西对外展示的一个窗口，“螺蛳粉热”便是其中的一个典型案例。螺蛳粉作为柳州的特色美食，从籍籍无名到在网络爆火，远销马来西亚、美国、加拿大和澳大利亚等多个国家，其工业化、产业化发展路径有目共睹，而其背后引发的饮食文化热潮也值得关注。

2020年，网友评选出中国对外输出的“新四大文化”——螺蛳粉、广场舞、中国网文、网购中国商品，其中螺蛳粉排名榜首，更引发了人们对于螺蛳粉的关注和热议。《光明日报》刊发评论文章《螺蛳粉输出的也是中华饮食文化》，认为螺蛳粉走俏海外体现的是“烟火人家的温婉”，展示的是“一道别具特色的文化景观”，“折射着中国波澜壮阔的减贫图景”，“体现了中国文化与世界越来越密切的融合”。

唯有文化自信，方能行稳致远。文化自信，是更基础、更广泛、更深厚的自信。螺蛳粉的热销及其背后引

In recent years, "Taste of Guangxi" has attracted more and more attention with its mysterious and rich characteristics. Its intangible cultural heritage diet as a representative has become a window for Guangxi's external display, and Luosifen fever is one of the typical cases. Luosifen has become the spiritual symbol and cultural pride of Liuzhou people and even Guangxi people.

发的饮食文化热潮，让原先没有多少存在感的广西特色饮食终于扬眉吐气了一回，螺蛳粉也成为柳州人乃至广西人的精神标识和文化骄傲。从这个角度而言，对其成功路径的研究，具有积极的现实意义。螺蛳粉的“出圈”，赋予了广西饮食前所未有的自信，也激励着更多广西特色美食再造新的辉煌。

山水相逢处，总有炊烟起。当下的中国早已度过了温饱阶段，美食带来的不仅是味觉享受，更是对地方文化的触碰与感知。伴随着国家崛起而成长起来的年轻一代，对自我文化的认同感越来越强烈，他们会“因为一种味道而爱上一座城”。而广西的饮食文化丰富多彩，鱼稻共生的田间美食、精彩纷呈的米粉江湖、深山密林的天然作料等，将如一个个微开的窗口，吸引着更多人驻足停留和探寻……原来，你是这样的广西！

At present, China has long passed the stage of feeding and clothing. Food brings not only taste enjoyment, but also touch and perception of local culture. The younger generation that grew up with the rise of the country has a stronger sense of identity with their own culture, and "falling in love with a city because of a taste" has become the emotional context and cognitive expression of young people. And Guangxi's colorful food culture will be like a slightly open window, attracting more people to stay and explore...

好山好水好食材 ◆孕育出不一样的"广西味道"

Picturesque mountains, clear waters, and high-quality ingredients ◆All of which nurture the unique "Guangxi Flavor"

后记
Epilogue

“非遗”是一个国家和民族历史文化成就的重要标志，是中华优秀传统文化的重要组成部分，是当代文旅融合的重要载体。习近平总书记关于“非遗”保护工作做出重要指示，并强调扎实做好“非遗”的系统性保护，推动中华文化更好走向世界。党的二十大也对“非遗”保护进行了部署。

“非遗”饮食与人们的日常生活息息相关，承载着千百年来地方文化基因、风土人情与饮食习惯，体现出中华文化的博大精深及中华历史文明的悠久深厚。

广西是多民族聚居的自治区，有壮、汉、瑶、苗、侗、仫佬、毛南、回、京、彝、水、仡佬等12个世居民族，形成了具有浓郁地方特色和民族特色的饮食体系。

本书选取广西具有代表性的16种“非遗”饮食，从溯源、技艺、传承三方面勾勒出广西“非遗”饮食发展图景，探讨“非遗”饮食的活态传承与创新发展路径，期望能为更多人了解广西的饮食文化提供帮助，为更多“非遗”饮食在带动就业、增加收入、延伸产业链、推动地方经济发展等方面提供参考。

全书的分工如下：宫斐负责全书文字的撰写和部分照片拍摄，刘娓、蒋岚岚协助进行传承人的采访、拍摄，李宁、谭艳红负责书中的英文翻译。特别鸣谢广西壮族自治区文旅厅非物质文化遗产处、广西非物质文化遗产保护中心工作人员的指导与支持；鸣谢“非遗”传承人甘保辉、黄永林、余富、黄学科、蒋福财、张泰华、龙付孙、黄天玲、韦洁群、石濡菲、周黎维、覃建良、莫桂坚、黄庆强、潘红华、周君翠，以及复记老友董事长黄宇等接受采访；鸣谢广西甘家食品有限公司、广西南宁复记老友餐饮企业管理有限公司、桂林市明桂米粉餐饮管理有限公司、南宁古源香餐饮管理有限公司、苍梧县六堡镇黑石山茶厂、广西丹泉酒业有限公司、广西梧州双钱实业有限公司、合浦公馆黄记食品有限公司、北海市君梦食品有限公司等企业的大力支持。

本书系广西教育科学“十四五”规划2022年度研学实践教育与劳动教育专项重点课题《广西非遗饮食活态传承与研学旅行开发互动融合研究》（编号：2022ZJY1675）的研究成果。

宫斐

2024年9月